Climate Change and Small Island States

Climate Change and Small Island States

Power, Knowledge and the South Pacific

Jon Barnett and John Campbell

First published in 2010 by Earthscan

For a full list of publications please contact:

Earthscan
2 Park Square, Milton Park, Abingdon, Oxfordshire OX14 4RN
711 Third Avenue, New York, NY 10017

First issued in paperback 2015

Earthscan is an imprint of the Taylor & Francis Group, an informa business

ISBN 13: 978-1-138-86696-6 (pbk)
ISBN 13: 978-1-8440-7494-5 (hbk)

Typeset by Kerrypress Limited
Cover design by Susanne Harris

A catalogue record for this book is available from the British Library

Library of Congress Cataloging-in-Publication Data

Barnett, Jon, 1971–
Climate change and small island states : power, knowledge, and the South Pacific / Jon Barnett and John Campbell.
       p. cm.
   Includes bibliographical references and index.
   ISBN 978-1-84407-494-5 (hardback)
1. Climatic changes–Government policy–Oceania. 2. Oceania–Climate. 3. Human
   ecology–Oceania. I. Campbell, John. II. Title.
   QC903.2.O3B37 2010
   363.738'74560995–dc22

                          2009031882

Contents

List of Figures and Tables

FIGURES

TABLES

Acknowledgements

Many people helped us in preparing this book. First and foremost among them are the thousands of people that we have had the pleasure to meet and learn from over our years of working in the region, most of whom we hope will appreciate our intentions and not disagree too much with our analysis. Some key people have helped us by providing access to materials, and answering specific questions, including in particular Patrina Dumaru, Taito Nakalevu, Diane McFadzien, Bec McNaught, Espen Ronneberg and Rossy Pulehotoa. We have also benefited greatly from specific inputs of various kinds from Gail Adams, Natasha Chamberlain, Ian Fry, Isabelle Kuntz, Simon Lambert, Josie Lee, Jo Lewin, Rebecca Marshall, Colette Mortreux, Max Oulten and Sophie Webber. Thanks to our families Heidi, Sabine and William, and Nittaya, Rebecca and Emily for tolerating our repeated physical and at times mental absences. Many thanks also to Lynda Johnston and Robyn Longhurst at the University of Waikato for their good humour and critical spirit. Finally, thanks also to Earthscan, and in particular Jonathan Sinclair Wilson for his support, and Claire Lamont, Camille Bramall and Gina Mance for their considerable patience and professionalism.

List of Acronyms and Abbreviations

ADB	Asian Development Bank
AIACC	Assessments of Impacts and Adaptations to Climate Change (project)
AOSIS	Alliance of Small Island States
APN	Asia Pacific Network for Global Change Research
AR4	Fourth Assessment Report (IPCC)
ASPEI	Association of South Pacific Environmental Institutions
BPOA	Barbados Programme of Action for the Sustainable Development of Small Island Developing States
CBD	Convention on Biological Diversity (UN)
CBDAMPIC	Capacity Building for the Development of Adaptation Measures in Pacific Island Countries
CCA	Climate Change Adaptation in Rural Communities of Fiji
CCS	Carbon Capture and Storage
CCSP	Climate Change Science Program
CDM	Clean Development Mechanism
CFC	chlorofluorocarbon
CGPS	continuous global positioning system
CHOGM	Commonwealth Heads of Government Meeting
CIDA	Canadian International Development Agency
COP	Conference of Parties
CROP	Council of Regional Organizations in the Pacific
CZM	Coastal Zone Management
EEZ	exclusive economic zone
EIA	environmental impact assessment
ENSO	El Niño Southern Oscillation
ESF	European Science Foundation
ESSP	Earth System Science Partnership
EVI	Environmental Vulnerability Index
FIELD	Foundation for International Environmental Law and Development
GCM	general circulation model

GDP	gross domestic product
GECAFS	Global Environmental Change and Food Systems (project)
GEC&HH	Global Environmental Change and Human Health (project)
GECHS	Global Environmental Change and Human Security (project)
GEF	Global Environmental Facility
GLP	Global Land Project
GNI	gross national income
GWSP	Global Water System Project
IAM	integrated assessment model
ICSU	International Council for Science
IFRC	International Federation of Red Cross and Red Crescent Societies
IGBP	International Geosphere-Biosphere Programme
IGCI	International Global Change Institute
IHDP	International Human Dimensions Programme on Global Environmental Change
IIED	International Institute for Environment and Development
IOC	Intergovernmental Oceanographic Commission
IOM	International Organization for Migration
IPCC	Intergovernmental Panel on Climate Change
ISSC	International Social Science Council
ITCZ	Intertropical Convergence Zone
IUCN	International Union for Conservation of Nature
LDCF	Least Developed Countries Fund
LOICZ	Land-Ocean Interactions in the Coastal Zone
LULUCF	Land Use, Land Use Change and Forestry
MIRAB	migration, remittances, aid and bureaucracy
NAPA	National Adaptation Plan of Action
NGO	non-governmental organization
NOAA	National Oceanic and Atmospheric Administration
NRC	US National Research Council
NTC	National Tidal Centre
NTFA	National Tidal Facility Australia
OECD	Organisation for Economic Co-operation and Development
OFCCP	Oceanic Fisheries and Climate Change Project
OPEC	Organization of Petroleum Exporting Countries
PACC	Pacific Adaptation to Climate Change (project)
PACESD	Pacific Centre for Environment and Sustainable Development
PCCP	Preparedness for Climate Change Programme
PIC	Pacific Island country
PICCAP	Pacific Islands Climate Change Assistance Programme
PIEPSAP	Pacific Islands Energy Policy and Strategic Action Planning (project)

PIFACC	Pacific Islands Framework for Action on Climate Change
PIFS	Pacific Islands Forum Secretariat
PI-GCOS	Pacific Islands Global Climate Observing System
PIGGAREP	Pacific Islands Greenhouse Gas Abatement through Renewable Energy Project
RSWG	Response Strategies Working Group
SBI	Subsidiary Body for Implementation
SBSTA	Subsidiary Body for Scientific and Technological Advice
SCCF	Special Climate Change Fund
SEAFRAME	Sea Level Fine Resolution Acoustic Measuring Equipment
SGP	Small Grants Programme
SIDS	small island developing states
SMIC	Study of Man's Impact on Climate
SOPAC	South Pacific Applied Geoscience Commission
SPC	Secretariat of the Pacific Community
SPCZ	South Pacific Convergence Zone
SPREP	Secretariat of the Pacific Regional Environment Programme
SPSLCM	South Pacific Sea Level and Climate Monitoring (project)
SRES	Special Report on Emissions Scenarios
START	SysTem for Analysis, Research and Training
STS	Science and Technology Studies
SPA	Strategic Priority for Adaptation
TAR	Third Assessment Report
TWAS	Third World Academy of Sciences
UNCBD	United Nations Convention on Biological Diversity
UNCED	United Nations Conference on Environment and Development
UNCTAD	United Nations Conference on Trade and Development
UNDP	United Nations Development Programme
UNEP	United Nations Environment Programme
UNFCCC	United Nations Framework Convention on Climate Change
UNHCR	United Nations High Commission for Refugees
UNITAR	United Nations Institute for Training and Research
UNU	United Nations University
USAID	United States Agency for International Development
USGCRP	United States Global Change Research Program
USP	University of the South Pacific
V&A	vulnerability and adaptation
WCED	World Commission on Environment and Development
WCP	World Climate Programme
WCRP	World Climate Research Programme
WG	Working Group (I, II or III)
WMO	World Meteorological Organization
WWF	World Wildlife Fund

1

The Trouble with Climate Change

Introduction

There is widespread consensus that climate change is extremely dangerous for small island developing states (SIDS). This has been well recognized in research and policy for over 20 years. Despite this, responses in the form of practical measures taken to implement adaptation have been few and piecemeal. Equally, there has been very little research that is oriented towards understanding how people living on islands can adapt to climate change in order to continue living lives that they value. These inadequate responses in research and policy become ever more acute as emissions of the gases that drive climate change continue to increase. This book describes and offers explanations for the difference between what is said about climate change in SIDS, and what is done.

We examine the intersections between science and policy in the response to global warming and identify the barriers that have restricted, and continue to restrict, responses that would benefit SIDS. In particular, we seek to explain why progress in implementing adaptation policies and strategies has been constrained. We draw almost exclusively on our experiences in research and policy processes relating to climate change in the Pacific Islands, and we focus in particular on research and policy relating to adaptation. Our principle argument is that the representation of climate change in SIDS is a discursive formation that limits understanding and action to address the interests of people living in islands. By 'discursive formation' we mean a system of statements that has regularity with respect to themes, objects and concepts, and the way they relate to each other (Foucault, 1989). Such formations often become naturalized and taken for granted as statements of truth and, while they can be subject to contestation and destabilization, they often hold hegemonic purchase such that other possibilities are precluded.

The discursive formation of the link between climate change and SIDS is easily identifiable. Its characteristic features include frequent reference to geographic objects (such as islands and coasts) with little if any recognition of the people that live in them. These island objects are seen as being all essentially alike, as are their human populations (if they are recognized at all). Statements about islands and climate change disproportionately focus on the environmental drivers of vulnerability – the changes in climate and sea levels and the fragility of island ecosystems – with little recognition of social factors that can enhance but can also significantly reduce the risk of damages arising from climate change. There is also a heavy emphasis on the problem (vulnerability) and little if any consideration of the means of and scope for local solutions (adaptation). There are assumptions about scale (large global forces literally and metaphorically drowning small islands); power (social-ecological drivers of vulnerability which overwhelm weak local systems); and knowledge (models and indexes are a *sine qua non* for decision making). The scope for and content of adaptation strategies is rarely considered, and, when it is, it is most often considered to be uncertain, with practice only possible after resolution of uncertainty delivered through application of the techniques of science. These regular statements have the effect of rendering climate change as an environmental fact against which actors can do little but suffer. They deny the agency of people at risk: to define the problem in their terms; to apply their own systems of knowledge; to implement the solutions that are appropriate to their needs and values and which accommodate uncertainty; and to make knowledge claims of equal value to those of science. This discursive formation has remained largely unchanged for more than a decade; a new way of representing Pacific Island capacities and potentials is required.

We argue that the discursive formation of climate change and SIDS is the product of power/knowledge in three ways. First, it is produced by networks of institutions (universities, the media, non-governmental organizations (NGOs), regional and global intergovernmental organizations) and policy regimes that gain or sustain power through the particular knowledge they (re)produce about climate change in islands. Second, it is a recent environmental manifestation of a longer discursive formation of islands as being sites of backwardness, insularity, constraint, fragility and weakness. This more long standing representation perpetuates early constructions of islands that were necessary to legitimate colonial interventions – at least in the minds of the colonizers. Colonial interventions included significant displacement of populations, at times through direct force, but more often through more subtle but no less devastating means. Third, we argue that the flow of power/knowledge is not unidirectional, but rather that there are: multiple sites of contest over knowledge claims; multiple objectives for which knowledge claims are produced; unintended consequences of knowledge claims; and changing configurations of knowledge/power which have the potential to transform the larger discursive formation of climate change in small islands.

This position we take on climate change in small islands is consistent with social constructivism in as much as we accept the notion that the production of knowledge is political, and that science is not necessarily, or cannot be unquestionably, the truth (Haas, 2004). We are concerned with how the perceived reality of climate change in islands comes into being, and particularly the ways in which what is known enables and constrains responses (Demeritt, 2006; Pettenger, 2007).

A problem with social constructivist approaches is that they give succour to climate sceptics, who seek to argue, by contesting science, that climate change is not a real problem, or not real enough to warrant significant policy responses (Demeritt, 2006). We are not climate sceptics. We do not doubt that climate change poses considerable dangers to the islands of the Pacific, as we explain later in this chapter. Indeed, our concern is that *not enough is being done* to address the problem, either in terms of reducing greenhouse gases or implementing adaptation responses. Our constructivist orientation is necessary, however, because the discursive formation of climate change in islands – to which we ourselves have contributed – constrains progress towards solutions. We show in this book the paradoxical role of science in this discursive formation. On the one hand its ability to identify the parameters of the problem and to mobilize responses has been and remains important; without it we would be unaware of the risks climate change poses to SIDS. On the other hand, the hegemony of natural science approaches to climate change and of modelling in particular, marginalizes other approaches to generating knowledge about climate change. In the absence of alternative forms of knowledge, or rather awareness that such knowledge can and does exist, certain actions, in particular those relating to adaptation, are impeded. The silence is sustained because as more scientific research is conducted in the name of reducing the uncertainties that are purported to impede action, new questions arise and further uncertainties can emerge. This further debilitates action (Demeritt, 2006).

We seek in this book not so much a complete recasting of what is known about climate change in SIDS, but rather a more nuanced representation, which acknowledges difference, embraces agency, and helps to disperse power in the implementation of responses. In this sense we seek to gently disassemble the current discursive formation by examining some of its specific dimensions and manifestations in order to open up space for new ways of understanding the problem, so that new solutions may follow.

To do this we examine the development of climate change science, particularly as it has (and has not) been applied in the Pacific Islands. We consider this science to cover the full range of disciplines ranging from understanding atmospheric processes through to the examination of the human dimensions of climate change. Informed by Science and Technology Studies (STS), we show that the formal international processes through which climate change science has been reproduced, and in particular the Intergovernmental Panel on Climate Change

(IPCC), have simultaneously appropriated and inculcated Pacific Island concerns. This arises from the dominance of 'front end' physical science, positivist and modelling methodologies, and the employment of social science informed almost entirely by the paradigms and methods of economics.

Similarly, the United Nations Framework Convention on Climate Change (UNFCCC), which is the major multilateral process intended to develop responses to climate change, and which is informed by the findings of the IPCC, has operated in a manner that makes it extremely difficult for small island countries to meaningfully participate. At one level, the intransigence of major greenhouse gas emitters has slowed the development of mitigation responses. At a deeper level, however, the absence of specific information about the social and cultural consequences of climate change in islands, and the forms of adaptation that are necessary, means that there is insufficient recognition of the real magnitude of climate dangers in the Pacific and so less impetus for emissions reductions. It also enables developed countries to argue for delaying assistance for adaptation on the grounds that information about adaptation possibilities is insufficient – an argument that suits the scientists whose research is oriented towards reducing uncertainty. The response of the UNFCCC and its financial mechanism (the Global Environmental Facility [GEF]) towards facilitating adaptation in the Pacific has been inadequate. Of course, a lack of information about climate change as a social and cultural problem in the South Pacific is only one of many reasons for the tardiness of the international community to reduce emissions and to help the Pacific Islands to adapt. However, it is an important one given that a large component of the little power these countries have in climate politics comes from the moral pressure they are able to exert (Paterson, 1996; Shibuya 1996).

There have been some regional initiatives in the area of adaptation, many supported by the GEF, but these are yet to yield appreciable benefits to the Pacific Islands. They include: a sea-level monitoring network; the development of an environmental vulnerability index; a short period of 'capacity building' of Pacific Island personnel; and a handful of 'adaptation' projects. While the first of these will provide little definitive information for several decades, the second remains incomplete, the third was truncated and the last is too little, piecemeal, and the lessons emerging from the best of them have not been widely recognized.

So, while climate change is considered to be a serious risk to the Pacific Islands (and we agree), there has been very little progress towards facilitating responses that would be of benefit to them. Given the delays in obtaining agreement to reduce greenhouse gas emissions we are particularly concerned with the paucity of action to implement adaptation in the Pacific Islands. This book seeks to explain the reasons for this slow progress.

The remainder of this chapter presents a brief introduction to the Pacific Islands, and a concise overview of the problem of climate change in the region.

Pacific Island countries

The Pacific Islands region comprises 22 countries and territories in the tropical Pacific Ocean (see Figure 1.1). As Table 1.1 shows, the countries vary considerably in terms of population and political status. As explained in Chapter 2, they also vary in terms of physical characteristics. There is considerable heterogeneity among the countries and, in several cases, within countries themselves.

Nine of the island states and territories in the region are fully independent, eight are dependent territories (either of France, the United States or New Zealand) and five are self-governing and constitutionally independent but with some form of association with either the United States or New Zealand.

Table 1.1 *Pacific Island countries and territories*

Country or Territory	Population (2008 est.)	Land area (km²)	Political status	Colonial connections[a]
MELANESIA				
Fiji Islands	839,324	18,272	Independent	United Kingdom
New Caledonia	246,614	18,576	Territory	France
Papua New Guinea	6,473,910	462,840	Independent	Australia
Solomon Islands	517,455	28,370	Independent	United Kingdom
Vanuatu	233,026	12,190	Independent	UK/France
	8,310,329	540,248		
MICRONESIA				
Fed. States of Micronesia	110,443	701	Free Assoc.	United States
Guam	178,930	541	Territory	United States
Kiribati	97,231	811	Independent	United Kingdom
Marshall Islands	53,236	181	Free Assoc.	United States
Nauru	10,163	21	Independent	United Kingdom
Northern Mariana Islands	62,969	457	Territory	United States
Palau	20,279	444	Free Assoc.	United States
	533,300	3,156		
POLYNESIA				
American Samoa	66,107	199	Territory	United States
Cook Islands	15,537	237	Free Assoc.	New Zealand
French Polynesia	263,267	3,521	Territory	France
Niue	1,549	259	Free Assoc.	New Zealand
Samoa	179,645	2,935	Independent	New Zealand
Tokelau	1,170	12	Territory	New Zealand
Tonga	102,724	650	Independent	United Kingdom
Tuvalu	9,729	26	Independent	United Kingdom
Wallis and Futuna	15,472	142	Territory	France
	655,200	7,981		
TOTAL	9,498,829	551,385		

[a] This column lists the colonial government immediately prior to independence for those countries that have attained that status or the current colonial government for those which have not. The table excludes Pitcairn Island. Source of data on Population and Land area: SPC (2008b)

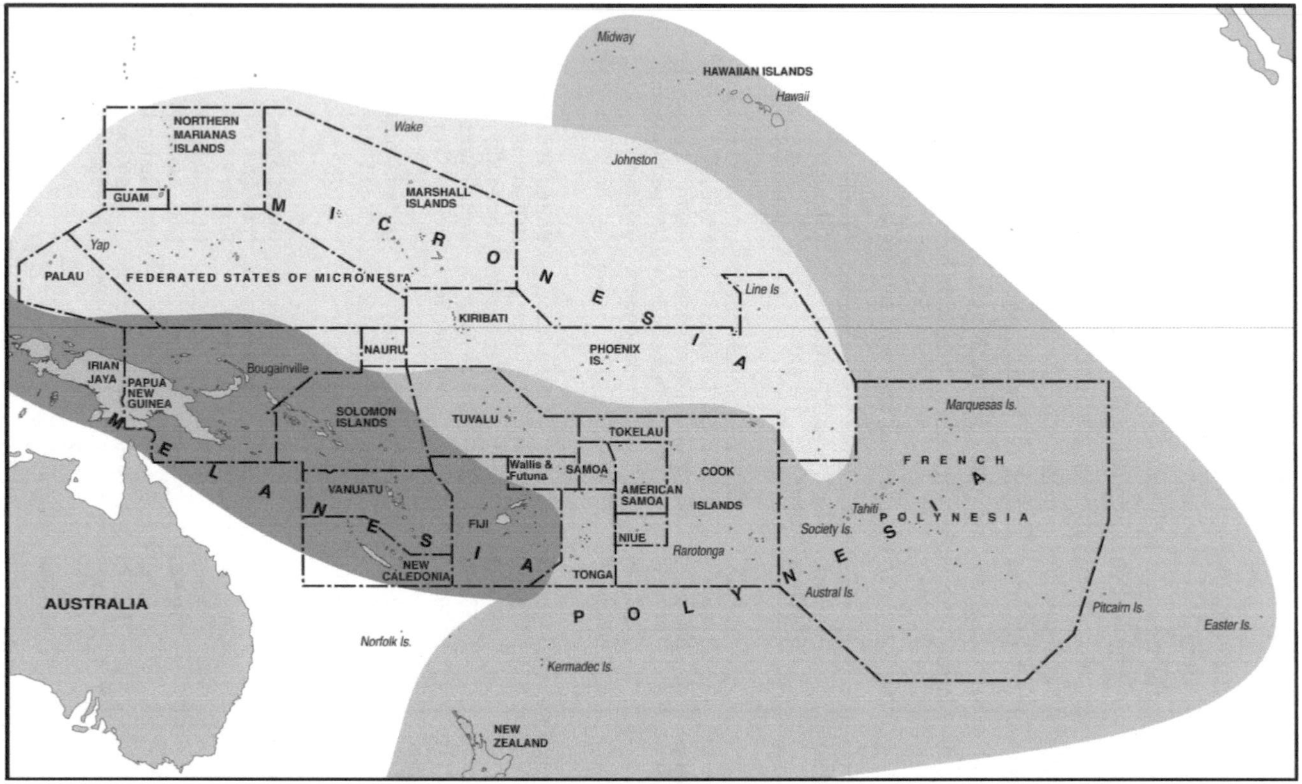

Figure 1.1 *The countries of the Pacific Islands region*

Note: The lines on the map are not territorial boundaries but serve to delineate the different states and territories. The map also shows the three Pacific Island subregions

The region is often divided into three subregions – Melanesia, Polynesia and Micronesia – although it should be recognized that these broad cultural categories obscure considerable internal variation. Melanesia is composed of the larger islands in the south-western part of the Pacific Ocean. Generally it is character-ized by considerable cultural diversity. For example, there are believed to be some 800 languages in Papua New Guinea, and over 100 languages are spoken among Vanuatu's 233,000 people. In comparison, Polynesia is relatively culturally homogenous, where, for example, there are many similarities among the approxi-mately 30 Polynesian languages, which are spoken in countries separated by thousands of kilometres of ocean. Micronesia, as its name suggests, is composed of a large number of small islands and in terms of cultural diversity lies between the other two regions. With the exceptions of Kiribati and Nauru, the Micronesian countries and territories have strong links to the United States.

Economic growth in the region has historically been slow, although since 2005 there has been a modest upswing in many countries (Commonwealth of Australia, 2008). Throughout the 1990s many economies in the region were stagnant, and some economies contracted. Gross national incomes (GNI) in the region range from US$15 million in Tuvalu to US$2.8 billion in Papua New Guinea in 2002 (ADB (Asian Development Bank), 2004). GNI per capita ranges from US$530 in Papua New Guinea to US$6,820 in Palau. Kiribati, Samoa, the Solomon Islands, Tuvalu and Vanuatu are classified by the United Nations as Least Developed Countries.

Key industries in the region include tourism, fishing licence fees and agricul-ture. Tourism accounts for 49 per cent of gross domestic product (GDP) in Palau, 47 per cent in the Cook Islands, 17 per cent in Vanuatu, 15 per cent in Kiribati and Tonga, and 13 per cent in Fiji (Commonwealth of Australia, 2006). Foreign-operated fishing boats pay licence fees to fish within the territorial waters of the equatorial countries and this provides them with a major income source. In Kiribati and Tuvalu fish licence fees account for over 40 per cent of GDP (ADB, 2002). With regard to agricultural exports, transport and shipping costs are too high for many countries to remain competitive in international markets. Across the region the share of primary industries contributes less than 20 per cent to GDP, with the one exception of Papua New Guinea where it is 33 per cent (Commonwealth of Australia, 2006).

Most of the smaller Polynesian and Micronesian economies are heavily dependent on aid and remittance income. Political and demographic ties to former colonial powers (in particular Australia, New Zealand, the United States) heavily shape the nature and extent of this income. In the Federated States of Micronesia, Kiribati, the Marshall Islands, Nauru, Niue and Tuvalu, aid accounts for at least one third of GDP. Remittances (money and goods sent from migrants living overseas – largely in Australia, New Zealand and the United States) are also important: in one of the most extreme cases remittances are thought to account for nearly 40 per cent of the GNP of Tonga (Browne and Mineshima, 2007).

Thus the income of many households in small islands comes from employment in the public service or on specific projects that are financed by aid, and/or from private sector activities which exist in part to supply the demand created by aid flows, and/or from remittances sent by family living abroad. The wealthier Polynesian and Micronesian countries are in part wealthier because many households benefit from aid and remittances in these ways. The majority of households in the larger Melanesian countries do not benefit from aid and remittances to the same degree, if at all. Nevertheless flows of money coming from distant lands are important when considering development in the region, and the potential effects of climate change on development.

Climate change and the Pacific Islands

The earth's climate is driven by the balance between incoming and outgoing radiation. The earth's surface absorbs approximately half of the incoming solar energy, causing heating. Some heat is re-emitted in the form of infra-red radiation, but most of it is blocked by a blanket of greenhouse gases (notably carbon dioxide, methane, nitrous oxides and halocarbons) that keeps the planet much warmer than it would otherwise be. Heating of the earth is greatest along the equator, and the broad patterns of climate and weather are determined by atmospheric and oceanic transportation of heat away from the equator and towards the poles.

Since the industrial revolution human activities such as land clearing and the burning of fossil fuels have increased the concentration of greenhouse gases in the atmosphere. Roughly 290 billion tonnes of carbon have been released into the atmosphere since 1751, half of which occurred after the mid 1970s (Marland et al, 2003). As a result, atmospheric concentrations of CO_2 (the most significant greenhouse gas) have increased by 30 per cent since 1750. These emissions have thickened the blanket of greenhouse gases, trapping more of the outgoing infra-red radiation, leading to warming of the atmosphere and the earth's land and ocean surfaces. This warming means that the process of redistributing heat from the equator to the poles is becoming more vigorous, leading to changes in atmospheric and oceanic circulations, weather patterns, and the hydrological cycle. The rates of warming now underway are considered to be much more rapid than at any time in the past 10,000 years. Indeed, 11 of the 12 years between 1995 and 2006 were the warmest on record (IPCC, 2007).

While knowledge about climate change has existed for some time (Weart, 2008) it was not until the late 1980s that it became an issue of international concern. In 1988 United Nations agencies formed the IPCC, an intergovernmental body, to assess the scientific evidence that such a process was happening, its likely effects and their significance and to consider ways of responding to it. The IPCC (some of the workings of which we discuss in Chapter 3) has produced four substantive three-volume Assessment Reports since 1990, each increasingly certain about the nature and causes of existing trends, and what future changes may emerge. The most recent of these is the Fourth Assessment Report (AR4), which indicates that it is very likely that increases in greenhouse gas concentrations have caused most of the increase in temperature observed in the second half of the 20th century, and that emissions of greenhouse gases will continue in the decades ahead, with resulting increases in temperature of between 1.1 and 6.4°C by the end of the century.

An important consequence of atmospheric warming is sea-level rise caused mostly to date by thermal expansion of warming surface waters, but increasingly in the future by melting of land-based ice in glaciers and in the polar regions (modelled projections range from 0.18m to 0.59m by 2100 (IPCC, 2007)). However, there remains a significant amount of uncertainty about projected sea-level rises and there is reason to believe that the above mentioned estimates may be conservative (Rahmstorf, 2007). Other anticipated changes include: increasing frequency and intensity of heavy precipitation events, droughts and heat waves; an increase in intense tropical cyclones; and increased frequency of high sea-level events. As we will see later, these extremes (together with geological hazards) are among the most commonly experienced by the majority of Pacific Island countries (PICs).

These changes in temperature, rainfall, sea level and extreme events pose myriad risks to ecosystems and the people that depend on them. The degree to which people are at risk from damages caused by changes in climate depends on: the extent to which they are dependent on ecosystems for their livelihoods (fishers are more dependent than soldiers, for example); the extent to which the ecosystems they depend on are sensitive to climate change (glaciers are assumed to be more sensitive than deserts, for example); and their capacity to adapt to these changes (people with insurance cover are better able to recover from an extreme event than those without insurance cover, for example). Capacity to adapt is a function of many factors, including: access to economic resources, technologies, information and skills; the degree of equity in a society; risk perception; and the quality of governance.

It is generally thought that income-poor people and low-income societies are more at risk from climate change than wealthy people and societies, and of these people, those living in the most sensitive ecosystems – such as people living in the Pacific Islands – are among the most at risk. However, there is something of a paradox with respect to assignations of vulnerability to the Pacific Islands: the general criteria for assuming vulnerability certainly raise grounds for concern, so that the Pacific Islands (and other SIDS) are the countries that people first think of in relation to climate change; yet they are amongst the places where the least is known about the ways that climate change will affect them and the ways in which these effects may be adapted to.

The measure of responsibility for the problem of climate change is an individual or group's share of the greenhouse gases currently in the atmosphere. These shares are typically allocated on a country basis, although, as Baer (2006) notes, between 1950 and 2000 the wealthiest 10 per cent of people in developed countries emitted 7.5 times more CO_2 than the poorest 10 per cent of people in developed countries, and 155 times more CO_2 than the poorest 10 per cent of people in developing countries. The United States is the country that is most responsible for climate change, having emitted 30.3 per cent of the CO_2 emitted between 1900 and 1999. The European Union countries were responsible for 22.1 per cent of emissions over the same period (Baumert and Kete, 2001). In terms of total CO_2 emissions, the five largest polluters in 2002 were the United States (24 per cent of global emissions), China, Russia, India and Japan (Marland et al, 2003). However, on a per capita basis people in Australia, the United States, Germany, Russia and the United Kingdom are the world's largest emitters of CO_2 (Turton, 2004). In general, then, the societies that are most responsible for the emissions of greenhouse gases are those that are least vulnerable because of the adaptive capacity conferred by the wealth they have generated largely through polluting forms of development.

It is commonly assumed that the PICs play a very small role in contributing to greenhouse gas emissions. A database on national emissions maintained by the World Resources Institute includes 11 of the 14 independent Pacific Island countries (WRI, 2008). Its data shows that in both absolute and per capita terms, emissions from the region are very low. Total emissions from the 11 countries account for around 0.04 per cent of the global total, and they comprise 7 of the lowest 12 emitters in the dataset. Indeed, the PICs contribute only 1.84 per cent of Oceania's total share (the bulk coming from Australia and New Zealand). In terms of per capita emissions too, many of the 11 Pacific countries rank lowly, with Kiribati second from the bottom, while all the others (except for Palau and Nauru) are well into the bottom half. In per capita terms Palau is the 24th largest emitter, and Nauru the 29th. These countries have quite small populations but reasonably high levels of consumption. Nauru was for a period a very wealthy country based on phosphate exports, although in recent years substantial economic contraction has likely resulted in substantial declines in emissions as well.

These data on emissions from the region only represent emissions of greenhouse gases and does not include the role of land use change and deforestation in contributing to atmospheric greenhouse gas concentrations. Here the situation does change a little. If this is included, the region's contribution increases to 0.37 per cent of the global share, the bulk of this resulting from logging in Papua New Guinea. If Papua New Guinea were excluded, the remaining ten countries contribute only 0.01 per cent of the emissions (i.e. their global share falls even more once land use is considered). The population of the region, then, produces far less than its hypothetical share of global emissions of greenhouse gases. Yet, as discussed in the following section, the region is highly at risk from the changes in climate arising from those emissions.

Table 1.2 *Greenhouse gas emissions from Pacific Island countries*

Country	Land use and forestry excluded		Land use and forestry included	
	MtCO$_2$	Tons CO$_2$ per person	MtCO$_2$	Tons CO$_2$ per person
Cook Islands	0	1.7	0	1.8
Fiji	2.5	3.1	2.7	3.3
Kiribati	0.1	0.6	0.1	0.6
Nauru	0.1	11.7	0.1	11.7
Niue	0	2.6	0	2.6
Palau	0.2	12.7	0.2	12.7
Papua New Guinea	9.1	1.7	155.1	29.3
Samoa	0.4	2.2	0.4	2.3
Solomon Islands	0.3	0.7	0.5	1.1
Tonga	0.2	2.3	0.2	2.3
Vanuatu	0.6	2.9	0.6	3

Source: Climate Analysis Indicators Tool (CAIT) Version 5.0. (Washington DC: World Resources Institute, 2008)

Climate change effects on Pacific Islands

The IPCC, AR4 (2007) has indicated with a very high degree of confidence that small islands are highly vulnerable to climate change. This assessment is based on the countries' high levels of exposure to sea-level rise and extreme events, and assumptions about their susceptibility to damage from those events and their limited adaptive capacity. Nevertheless, there has been relatively little research into the effects of climate change on SIDS. Indeed, the IPCC commented in its AR4 (Mimura et al, 2007) that there has been a decline in the number of studies of climate change in small islands, after an initial earlier burst based on simple scenarios and methods. The IPCC lamented that it could refer to very few new robust studies in the period between the Third (2001) and Fourth (2007) Assessment Reports. More disturbing is the fact that a similar observation was made at the end of the small islands chapter in Third Assessment Report in 2001, which observed that 'insufficient resources are being allocated to relevant empirical research and observation in small islands' (Nurse and Sem, 2001, p870). Nevertheless, we can outline in broad terms some of the effects of climate change that are likely to be experienced in the Pacific Islands.

There are five large-scale changes in climate that are likely to affect the Pacific Islands region. First, air temperature is projected to increase by between 0.99°C and 3.11°C by the year 2099 relative to 1961–1990 average temperatures (Ruosteenoja et al, 2003). These estimates may be conservative given that there have been decadal increases in the region of between 0.3°C and 0.5°C since the

1970s (Salinger, 2001). It is also expected that there will be sea-surface temperature increases, which may have severe implications for coral ecosystems. Second, rainfall events will be more intense and possibly less frequent, implying both more flooding and drought events (Jones et al, 1999). In the already wet summer period more rainfall is expected, whilst there may be less rainfall in the already dry winter months. This will have implications for sustaining crops throughout the year as few crops are currently irrigated. Third, as discussed earlier, it is expected that there will be increases in sea levels.

There are also likely to be changes in regional climate systems. Of particular importance is the El Niño Southern Oscillation (ENSO). El Niño years bring drought to most of the Pacific Islands, and the 1997/98 El Niño caused widespread drought and food shortages: agricultural losses in Fiji were valued at US$65 million, and some 260,000 people in Papua New Guinea were placed in a life threatening condition due to depleted food supply (WMO, 1999). There is uncertainty about the effect of climate change on ENSO, but since the 1970s there have been more frequent and intense El Niño events. Finally, tropical storms may also become more intense in the future. In many Pacific Islands cyclones are a cause of mortality and injury, and they can cause massive financial losses. Cyclone Ofa which struck Samoa in 1990 caused over US$100 million in damage, as did cyclone Kina which struck Fiji in 1993 (Campbell, 1997; Olsthoorn et al, 1999). Cyclone Heta which struck Niue in 2004 destroyed the national hospital, national museum and the bulk fuel storage tanks (Government of Niue, 2004). In 2007, flooding caused by Cyclone Guba caused 149 deaths in the Oro province of Papua New Guinea and 58,000 people required food relief and other assistance (IFRC, 2008). As well as wind damage and damage from increased rainfall and flooding, cyclones induce storm surges which can reach up to six metres in height, and, in the case of cyclone Heta in Niue, waves in excess of 30 metres in height were observed.

It is the possibility of increases in the frequency and/or intensity of such hazards, rather than changes in mean conditions, that pose the most immediate danger to Pacific Islands. A critical factor for social-ecological systems in the region will be the extent to which return times between extreme events decrease, which would decrease the ability of systems to recover. In the longer term these changes in extremes will be compounded by changes in mean sea level, temperature and rainfall. Both these changes in extremes and mean conditions pose dangers to the Pacific Islands through their impacts on water reources, agriculture, fisheries, health, economic development, settlements and food security.

The IPCC, AR4 (Mimura et al, 2007, p689) considers with very high confidence that water resources in small islands are likely 'to be seriously compromised' by climate change. It is possible to identify several pathways through which this may come about, as well as several different secondary effects. While rainfall reductions may be the most significant cause of reduced fresh water availability, other factors are also important. Coastal erosion can reduce the volume of the

Ghyben-Hertzberg lens that is a critical natural freshwater store on those islands that are atolls. Similarly, salinization as a result of wave action and storm surge can degrade freshwater supplies for coastal communities and may damage crops in coastal areas. River flooding on larger islands can cause temporary damage to water supply systems in the event of tropical cyclones. In turn a reduced, degraded or less reliable fresh water supply can have adverse effects on other sectors including public health, agriculture, tourism (which is highly water dependent) and hydro-electricity production.

Agricultural production in the Pacific Islands is likely to be adversely affected by climate change through loss of coastal lands, increased contamination of groundwater and estuaries by saltwater incursion, and losses due to cyclones and storm surges, heat stress and drought. The increased risk of flooding in river catchments also threatens food production. For example severe flooding of the Wainibuka and Rewa Rivers in Fiji in April 2004 caused damages to between 50 and 70 per cent of crops (Government of Fiji, 2004). If agricultural production is reduced then some countries may face reduced earnings from agricultural exports, while at the same time experiencing reductions in subsistence (and market) food availability. Access to fisheries is also at risk from changes in fish habitats, migration patterns and damages to fishing related infrastructure. This has serious implications for food security given that many Pacific Island communities are highly dependent upon the marine environment as a source of protein.

The health of Pacific Island people may also be adversely affected through extensions in the range of mosquitoes that carry malaria and dengue fever, as the factors that determine the range of these insects are influenced by climate. There are also demonstrated positive associations between temperature increases and diarrhoea, and between warmer sea-surface temperatures and ciguatera outbreaks (Hales et al, 1999; Singh et al, 2001). Heat stress, and increased injuries and deaths from extreme events are also likely to result.

There may be some significant costs to island economies resulting from climate change. For example, the World Bank (2000) estimates that by 2050 damages from climate change could cost Tarawa atoll in Kiribati $8–16 million, or 17–34 per cent of current GDP. Another study estimates that the economic impacts of climate change on Pacific Island economies may be 'so profound that they dwarf any strategic issue currently confronting a major peacetime economy' (Hoegh-Guldberg et al, 2000, p4). Impacts on tourism have yet to be seriously examined, but it is believed that the industry may be affected through the loss of beaches, and indirectly through milder winters in traditional markets, reducing the motivation to take vacations abroad (Becken, 2005). Extreme events will also be increasingly costly for tourism infrastructure. So, some of the region's main forms of income generation – agriculture, fisheries and tourism – are likely to be adversely effected by climate change, and it then follows that employment in these sectors may also suffer. The impacts of climate change on these key sectors may also have other important secondary effects; for instance, not only farmers' livelihoods are at risk from climate change, but also those whose livelihoods depend on agricultural production, such as input, transport, information and credit suppliers. Impacts in one sector may in turn affect others; for example, declining incomes from agriculture may cause migration to urban areas, increasing urban poverty and placing increasing demand on urban services.

Many Pacific Islands are composed entirely of coastal zones in the sense that all of the land surface is affected by sea. Accordingly, the likely ramifications of sea-level rise and increased storm intensity are severe. A large proportion of Pacific Island communities, both rural and urban, are in coastal locations, and coastal resources play an important role in the subsistence and cash economies. Sea-level rise will aggravate coastal problems such as inundation, storm-surge impacts during tropical cyclones, erosion, and salinization of soils and ground water. As we will discuss later, many communities may find their current locations untenable and be required to relocate. With all but a few exceptions, urban areas in PICs are located on the coast, many of them being the former administrative centres and ports of colonial governments. Moreover, most PICs have only one major urban centre (Papua New Guinea and Fiji are the exceptions) and most of many countries' critical infrastructure is in these urban localities. Coral reef systems may also be at risk from rapid changes in sea-surface temperature, which cause coral bleaching (Reaser et al, 2000). Rising concentrations of CO_2 in the oceans may also retard the ability of reefs to grow in step with sea-level rise (Kleypas et al, 1999). Bleaching of reefs causes erosion of shorelines through changes in the supply of sediment. For atolls, which usually rise no more than a couple of metres above sea level, and whose morphology is dependent on reefs, the risks posed by climate change are particularly acute.

Through its potential impacts on agricultural production and fisheries, climate change may undermine the local availability of food. Through its impacts on employment and wages, it may also undermine the ability of people and societies to purchase food. Thus climate change puts at risk food security – that is the need for people in the islands to have access to sufficient, safe and nutritious food at all times. Indeed, a number of PICs are currently facing increasing problems of food security. This reflects the increasing proportion of populations in urban areas, perishability of some (but not all) traditional staples, land degradation and reduced yields, and population growth (Sharma, 2006; FAO, 2008). Indeed the complexity of climate change effect pathways is illustrated by the links among reduced food security, reduced fresh water supply and/or quality and increased spatial range of disease vectors, which when combined have significant implications for the health of Pacific Island communities.

In the discursive formation of climate change in SIDS, sea-level rise is the focus of attention, particularly as it may impact on atolls. Yet the Pacific Islands may be affected in a broad range of ways, some of which will act independently from sea-level rise, and many of which will be compounded by sea-level rise. These effects may be multiple, cumulative and/or overlapping. Among other things, developing measures to ameliorate their negative impacts will require careful consideration of the linkages among the various effects in order to identify appropriate points of intervention.

Responding to change

Pacific Island communities have been confronted with a considerable degree of externally imposed change over the past two centuries or so. Colonialism saw major transformation of traditional social, political, economic, religious and resource-management systems. Large parts of the region were theatres of battle during the Second World War, it was a strategic region during the Cold War, and it has since been buffeted by the policy demands associated with various waves of development thinking. By and large most communities adapted to these and other changes remarkably well. They have developed strategies to appropriate that which was seen as useful, a process Sahlins (2000) has called 'the indigenization of modernity', where traditional systems incorporate the most suitable and useful aspects of colonial and contemporary global society. Communities have also dem-onstrated resilience in the ways that they have managed those changes that have been disruptive.

There are two basic responses to global warming: mitigation (which is the term chosen by the international community to describe actions that seek to reduce the rate at which greenhouse gases are emitted into the troposphere), and adaptation, which refers to actions that people may take in order to cope with the effects of global warming. Mitigation is not the main theme of this book. As is indicated earlier, PICs are generally very small emitters of greenhouses gases, so that while there are very real benefits to be gained by reducing emissions from these countries, and it is important that they demonstrate that they are doing their share of the work in reducing emissions, it is still the case that, even on a per capita basis, they contribute very little to the causes of climate change. Neverthe-less, we are concerned with the risks climate change poses to PICs. The magnitude of those risks is a function of concentrations of greenhouse gases: the higher the concentrations, the faster and greater the rate of climate change, and the more likely it is that adaptation to those changes will be inadequate.

Recent research that projects future concentrations of greenhouse gases based on emissions trends since 2000, coupled with evidence of emissions reductions in practice, shows that it is optimistic to assume concentrations of greenhouse gases in the atmosphere will stabilize at 450 parts per million volume CO_2 equivalent, or that global warming will be constrained to 2°C above pre-industrial levels (Anderson and Bows, 2008). Atmospheric concentrations of 650ppmv CO_2 equivalent are likely, resulting in mean warming of 4°C or higher, with serious implications in terms of impacts (Parry et al, 2008). This is very significant for SIDS, given that at even 2°C localized warming is likely to cause annual coral bleaching (Donner et al, 2005).

A much greater effort towards mitigation is therefore clearly required. This includes those developed countries that are listed in Annex-1 of the Kyoto Protocol, but also those large polluters among the non-Annex-1 countries such as Brazil, China, India and Indonesia. While we agree that developing countries have legitimate grounds for expecting the industrialized nations, who are responsible for the great majority of existing additional greenhouse gases, to take the lead, we are concerned that unless the large developing country emitters do not reduce their emissions SIDS will suffer inordinately. Thus, parts of the South stand to lose for as long as major emitters allow the North–South stand-off in the climate regime to impede actions to reduce global emissions. This said, existing concentrations of greenhouse gases mean that there is already a commitment to warming even if all emissions were now to cease. However, as discussed above, a significant amount of further warming is likely given recent emissions trends and projected future emissions. Despite the best efforts of many, including the SIDS (discussed below), the UNFCCC and the Kyoto Protocol has thus far been unable to reduce emissions to anything like the level required. For example, the Kyoto Protocol committed just the Annex-1 countries to a 5 per cent reduction of the main greenhouse gases below 1990 levels, but will do almost nothing to slow the rate of climate change given that at least a 75 per cent reduction of global emissions below current levels is necessary to avoid 'dangerous' climate change (Stern, 2007). Accordingly, adaptation is becoming increasingly urgent as a climate change response in the countries of the Pacific Island region.

Adaptation means adjustments to reduce vulnerability to observed and expected changes in climate. Its practice entails one or more of three things: reducing the sensitivity of the system exposed to climate change (for example building irrigation systems in areas where rain-fed agriculture may be less tenable due to increasing variability); altering the exposure of a system or group to climate change (for example by relocating communities living in flood-prone areas); and increasing adaptive capacity (for example subsidizing disaster insurance to households that can't afford it) (Adger et al, 2005). The aforementioned ability of Pacific Island communities to manage past changes with some success suggests that there is reason to think that they have some capacity to adapt to climate change. In practice, the degree to which this capacity translates into effective responses depends very much on the nature, magnitude and speed of change, and the specific attributes required to adapt to the particular effects of climate change in any given place. These issues are poorly understood and under-researched.

The concept of adaptation has its origins in evolutionary biology, where successful adaptation is seen as the basis for natural selection. This is the idea that organisms can modify themselves or their behaviour in order to make best use of environmental conditions. Adaptation implies long-term processes such that the changes become accepted as biologically or behaviourally normal. In a human social or cultural sense the notion of adaptation becomes a little more problematic. From a functionalist perspective, many social or cultural patterns may be attributed to adaptation to environmental conditions. However, the reasons for the existence of many such elements may have other explanations that are rooted in the social, political and cultural structures that exist within communities. For example, it may be claimed that the *yavu* or raised house mound in Fiji is an adaptation to flooding which is often experienced during tropical cyclone events. The *yavu* also stands as the core of one's family identity and connection with place, a fundamental relationship in many Pacific Island communities. The *yavu* also has a political role: the higher one's status, the higher the house mound within the village. From these perspectives, then, the *yavu* is probably an incidental adaptation that does help keep people and their possessions above water during flood events, but this function of the *yavu* cannot be understood independently from its social and cultural meaning.

The hints of environmental determinism in the functionalist approach to adaptation aside, the important point here is that adaptations will not be implemented, nor be successful, unless they are consistent with the social and cultural mores particular to the community in which adaptation takes place. Put another way, adaptation activities have to be aligned with what people consider to be their needs, rights and values, otherwise implementation of adaptation will fail. It is therefore important that adaptation strategies are not imposed by outsiders, and that local communities are carefully and deliberately involved and empowered in decision making about and implementation of adaptation activities.

About this book

This book does two things. First, it provides information about the problems climate change poses for the Pacific Islands, and what is being done and what could be done in response. Second, it provides analysis of the ways in which climate change in the Pacific Islands is represented, and the ways these representations constrain responses.

To do this, the next chapter outlines the changes in the relationships between people and their environment in the Pacific Islands. This context is necessary to set the problem of climate change in the region in its appropriate temporal and spatial context. It shows that Pacific Island communities have historically been capable of adapting to change, but that these capabilities are rarely recognized in outsiders' representations of the region.

Chapter 3 discusses the way that knowledge about climate change, in particular as it affects the Pacific Islands, is produced. This chapter also considers the networks of power that are bound up in knowledge of climate change, and the implications of that knowledge and alternative forms of knowledge for policy and practice. It argues that there are significant and avoidable limitations in knowledge about climate change in the region, and that this is a key reason for the hitherto minimal and sub-optimal responses to the problem.

Chapter 4 explores the initiatives undertaken in the Pacific Islands region to produce climate knowledge and to inform the policy process. It shows that integrated assessment models (IAMs) have become the dominant form of such knowledge (re)production and this has emerged at the expense of other possible knowledge pathways, including local perspectives and traditional knowledge systems. As a result projects to promote adaptation have tended to be top-down and neglect the needs and aspirations of local communities.

Chapter 5 then explains the ways in which the global climate change policy process manifests itself in the Pacific Islands, and the way the Pacific Islands themselves engage in the policy process. It argues that the Pacific Islands are not without 'climate power', but nor are they 'climate powerful'. The chapter also considers the position of powerful metropolitan countries with respect to the problem of climate change in the Pacific Islands, and it examines the resourcing barriers to the Pacific Islands increasing their power in global and regional climate change negotiations.

Chapter 6 examines the projects and programmes that have been implemented in the Pacific Islands in response to climate change. It shows that climate projects implemented in the region are few, largely focused on reducing greenhouse emissions, ad hoc in nature and lack sustained commitment from donors. The chapter argues that there is something of a mismatch between that which Pacific Islanders consider to be desirable climate developments and that which global and metropolitan donors are willing to supply. A number of reasons are identified for this, including the power of donors, the power of the 'science' that donors can bring to bear in the context of the uncertainty of knowledge about climate impacts, and the multiplicity of actors and the absence of effective coordination mechanisms.

Chapter 7 grounds our critiques of the science-policy process in two case studies. The first, the AusAid funded South Pacific Sea Level and Climate Monitoring Project, is a science-heavy project which has been informally criticized by many as being a costly investment to produce, rather than reduce, uncertainty about climate impacts. The second project, the New Zealand funded Environmental Vulnerability Index, is less ideologically driven, but is nevertheless an expensive, epistemologically impossible exercise, based on abstract and unreliable data. Both examples show that the nature and integrity of the science on climate change in the region is compromised by networks of institutions that gain power through the particular knowledge they (re)produce about climate change in Pacific Islands.

In Chapter 8 we outline a number of discursive representations of PICs that have emerged around the issue of global warming. These are widely reported in media representations of the region and are often supported by concerned groups such as international environmental organizations. We argue that these representations have portrayed the Pacific Islands in ways that render the possibility of responding to global warming seemingly pointless, and deny the resilience and agency that Pacific Island communities have and which could be useful elements of an adaptation response.

Chapter 9 concludes the book, summarizing our evidence and arguments, and proposing some ways forward in terms of both representing the problem of climate change in the Pacific Islands, and improving responses.

2

Environment and Development in Pacific Islands

Introduction

The 'island' has become an icon in climate change discourses, frequently mentioned for its acute vulnerability to sea-level rise, and in more nuanced accounts to extreme events. One problem with such discourses is that they mask the heterogeneity of island environments and of their social systems. The effects of climate change on islands and the communities that live on them are likely to be highly differentiated: not all places will experience the same changes; where changes are similar the magnitude and timing of them will likely differ; the sensitivity of ecological and social processes to changes differs from place to place; the capacity of social systems to adapt to these changes is also not homogenous; and the significance of changes to social systems will also differ (different communities value things differently). Within the Pacific Islands there is such heterogeneity in social and ecological systems across islands, and indeed in the case of many parts of Melanesia, within single islands, that simple claims about 'island vulnerability' are empirically impossible to sustain. This is important to recognize, not least because it means that 'one size fits all' approaches to adaptation are unlikely to be successful in any given place.

In this chapter, we illustrate the considerable variety of islands, in terms of their geomorphology, climate and biota, and human uses of these elements. The chapter also outlines changes in the relationships between people and their environment in the Pacific Island region over time. In the jargon of climate change impact assessment, this chapter provides baseline information on the environment and society of Pacific Island countries (PICs). The broad factors that determine the exposure of islands to climate risks, their sensitivity to those risks, and their capacity to adapt to them are also identified. These factors shape vulnerability to climate change, which is the subject of the concluding section.

We have struggled in preparing this chapter, and indeed the whole book, with the dualism implicit in the presentation of people and nature as separate entities.

We do not see them as such. However, given the pervasiveness of modern think-ing in which society is so forcefully and repeatedly separated from nature (Leiss, 1972; Plumwood, 1993), finding a new and consistent language to describe places as hybrid socio-natural entities where both the 'social' and 'natural' are so intertwined as to be analytically inseparable is a philosophical project beyond the scope of this book (Swyngedouw, 1999; White, 2006). We will return later (in Chapter 3) to the problem of the separation of society and environment in climate change science; suffice it to say now that we do not accept that nature is divisible from the social, rather, 'the environment *is* economic, and it is also social and political life and cultural sustenance' (Banks, 2002, p42). It is our experience that because many Pacific Island communities do not see themselves as ontologically separate from nature in any possible way, actual or impending changes in 'environmental' elements mean quite different things to local people than they do to outsiders. This is expressed most effectively by Ravuvu, who discusses the notion of the *vanua* in relation to a village in Fiji:

> *The people of Nakorosule wherever they are and in whatever work they are involved are often reminded by their elders not to forget the Vanua, meaning the land and the social system and the dela ni yavu, one's house site back in the village … The Vanua in terms of the dela ni yavu is the physical embodiment of one's identity and belonging. (Ravuvu, 1988, p6)*
>
> *The people of Nakorosule cannot live without their physical embodi-ment in terms of their land, upon which survival of individuals and groups depends. It provides nourishment, shelter and protection, as well as a source of security and the material basis for identity and belonging. Land in this sense is thus an extension of the self, and conversely the people are an extension of the land. (Ravuvu, 1988, p7)*

Elsewhere he writes that the concept of *vanua* represents the totality of a Fijian community (Ravuvu, 1987). Such relations are found throughout the region and the reader is urged to keep these concerns in mind when reading this chapter.

Pacific Island environments

The Pacific Islands are widely varied in terms of their ecological characteristics. A fairly basic distinction is made between the oceanic islands, which are found east of the andesite line (which runs along the boundary between the oceanic and continental plates in the western Pacific and is part of the 'ring of fire'), and the 'continental' type islands that lie to its west (Thomas, 1963). Nunn (1994) prefers to use the terms 'intra-plate islands' and 'plate boundary islands' respectively, which tend to be more indicative of their respective geological origins.

Table 2.1 attempts to summarize the physical characteristics of the 21 Pacific Island political entities. Some countries, such as Fiji, are not composed solely of one type of island. Given that oceanic islands are often found in linear arcs or clustered archipelagos, a range from volcanic high islands through to atolls may be found within the one country: for example Papua New Guinea, which is by far the largest country in the region, is composed mostly of continental type islands, but also has numerous small islands off its coast and even atolls to the north-east of Bougainville. As the table shows, there are four countries or territories that are composed only of low-lying atolls: Tuvalu, Kiribati (although it does contain a raised limestone island, Banaba), the Marshall Islands and Tokelau.

Table 2.1 *Physical characteristics of Pacific Island countries and territories*

Country	Land area (km²)	Highest elevation (m)	Main island type
American Samoa	199	964	V
Cook Islands	237	652	V & A
Federated States of Micronesia	701	791	V & A
Fiji Islands	18,272	1,324	P-B
French Polynesia	3,521	2,241	V & A
Guam	541	406	V
Kiribati	811	81	A
Marshall Islands	181	10	A
Nauru	21	61	RL
New Caledonia	18,576	1,628	P-B
Niue	259	68	RL
Northern Mariana Islands	457	965	V
Palau	444	242	V
Papua New Guinea	462,840	4,509	P-B
Samoa	2,935	1,857	V
Solomon Islands	28,370	2,447	P-B
Tokelau	12	5	A
Tonga	650	1,033	V
Tuvalu	26	5	A
Vanuatu	12,190	1,879	P-B
Wallis and Futuna	142	765	V

Note: A = atoll(s); P-B = plate-boundary island(s); RL = raised limestone island(s); V = volcanic high island(s).

The plate-boundary islands are those found in the westernmost Pacific and which are formed by subduction of the denser oceanic Pacific Plate under the Indo-Australian Plate. These islands, which coincide with the cultural region of Melanesia (not including New Zealand), tend to be much larger, have higher elevations, relatively well developed soils and a much broader range of environments including large river catchments and flood plains. The plate-boundary island groups also include numerous associated offshore small islands, including volcanic islands and atolls.

Figure 2.1 shows the main types of intra-plate islands which are found in the eastern or central Pacific, and which range from large islands of volcanic origin (some volcanoes such as Mauna Loa in Hawaii and Matavanu in Samoa are still active) through to low-lying atolls, with a range of island sizes, heights and shapes between these extremes. Intra-plate islands are typically formed above hotspots in the earth's mantle. A series of islands may be formed over the one hotspot. As oceanic plate movement takes them away from the hotspot, subsidence and processes of erosion cause them to diminish in elevation and area. Typically, oceanic island arcs are formed, with the islands furthest from the hotspot being the lowest and smallest. An example of such a chain is the Hawaiian islands. To the east, situated above the hotspot is the 'big island' of Hawaii where volcanic action is continuing to create new land. As one moves along the Hawaiian island chain the islands become older, smaller and lower. Some 2500km to the west the chain ends with Kure atoll whose origins were some 30 million years ago (Macdonald, 1983).

After the islands are formed, fringing coral reefs develop around their coastlines. These reefs are attached to the coast and there is no lagoon separating them from the land. Coral reefs are composed of the skeletons of polyps and build up over time as generation after generation of the polyps die. With erosion and subsidence of the land the reef can become separated from it to become a barrier reef. Eventually the island may subside completely, leaving only the reef as the

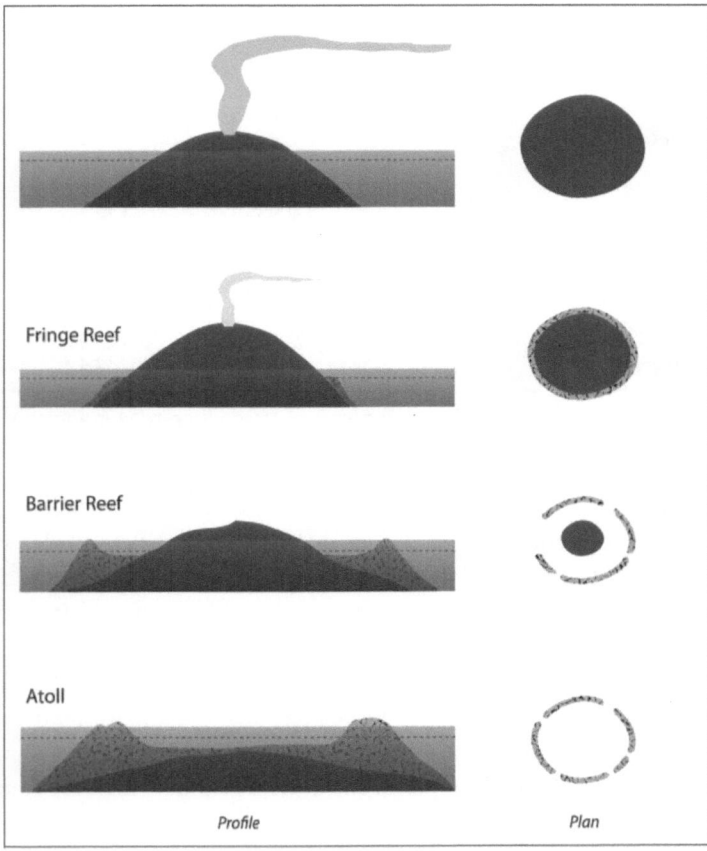

Profile   Plan

Figure 2.1 *Types of intra-plate or oceanic island*

corals continue to grow, providing the rate of subsidence is not too rapid. Atolls are formed under these circumstances, when the volcanic core subsides below sea level and coral continues to build on what were originally barrier reefs. As coral grows below sea level, atolls are formed by coral material being deposited on the reefs by wave action, building an above sea-level land area upon the reef in the form of small islets commonly called *motu*. Ironically, tropical cyclones which cause great devastation on atolls also play an important role in building them. Typically such atoll building takes place in areas facing prevailing trade winds (and waves) and these sides of atolls tend to have larger land areas than those on the leeward sides. Much of the global concern about climate change impacts is based on the possible plight of atoll dwellers who have no high land to use as refuge from rising sea levels or extreme tidal, high-wave or storm-surge events, or who have no spare land to move to in the event that an islet is inundated. In addition there are raised limestone islands, or raised atolls (sometimes called *makatea*) such as Niue and Nauru, which, as their name implies, are atolls that have been stranded above sea level either through geological processes or previous periods of sea-level change. Table 2.2 outlines some of the characteristics of the different island types and their exposure to the risks arising from climate change.

Table 2.2 *Types of island in the Pacific region and their exposure to risks arising from climate change*

Island type	Exposure to climate risks
Plate-boundary islands Large High elevations High biodiversity Well developed soils River flood plains Orographic rainfall	Located in the western Pacific these islands are exposed to droughts. River flooding is more likely to be a problem than in other island types. Exposed to cyclones, which cause damage to coastal areas and catchments. In Papua New Guinea high elevations expose areas to frost (extreme during El Niño), however highlands in Papua New Guinea are free from tropical cyclones. Coral reefs are exposed to bleaching events. Most major settlements are on the coast and exposed to storm damage and sea-level rise.
Intra-plate (Oceanic) islands **Volcanic high islands** Steep slopes Different stages of erosion Barrier reefs Relatively small land area Less well developed river systems Orographic rainfall	Because of size few areas are not exposed to tropical cyclones, which cause most damage in coastal areas and catchments. Streams and rivers are subject to flash flooding. Most islands are exposed to drought. Barrier reefs may ameliorate storm surge and tsunami. Coastal areas are the most densely populated and exposed to storm damage and sea-level rise. Localized freshwater scarcity is possible in dry spells. Coral reefs are exposed to bleaching events.

Table 2.2 *(continued)*

Island type	Exposure to climate risks
Atolls	
Very small land areas	Exposed to storm surge, 'king' tides and high
Very low elevations	waves, although exposure to cyclones is much less
No or minimal soil	frequent than in islands to the west and south.
Small islets surround a lagoon	Flooding arises from high sea-level episodes.
Shore platform on windward side	Exposed to fresh water shortages and drought.
Larger islets on windward side	Freshwater limitations may lead to health prob-
No surface (fresh) water	lems. Coral reefs are exposed to bleaching events.
Ghyben Herzberg (freshwater) lens	All settlements are highly exposed to sea-level rise.
Convectional rainfall	
Raised limestone islands	
Steep outer slopes	Depending on height may be exposed to storm
Concave inner basin	surges and wave damage during cyclones and
Sharp karst topography	storms. Exposed to fresh water shortages and
Narrow coastal plains	drought. Fresh water problems may lead to health
No surface water	problems. Flooding is extremely rare. Coral reefs
No or minimal soil	are exposed to bleaching events. Settlements are
	not exposed to sea-level rise.

This discussion about island geology highlights that islands are highly dynamic environments. Their shapes, elevation and even location are constantly changing. There are two important implications of this when considering climate change. First, the people that inhabit islands live with these changes and in most cases have adapted to them and continue to do so. Many environmental changes are slow, such as occurs with processes of lifting (as is the case in Niue), subsiding (as is the case in Maui), and tilting (as is the case in Tongatapu), allowing plenty of time for gradual adjustment. But there are also rapid and intense changes, such as occurs with volcanic events and cyclones, causing damage to ecosystems upon which people depend, property and in some cases loss of life. Rarely, however, do such events represent 'tipping points' in which social practices are radically changed. Environmental change is a feature of island life, and people living in islands tend to understand it as such and have developed capacities to adjust in response to such changes. These capacities, which we discuss in more detail later, may be stretched by the more frequent and intense changes that seem likely as a consequence of climate change, so much so that adaptation may necessitate critical changes in social systems. Adaptive capacity may also be undermined or constrained by non-climate related social and economic changes such as increasing urbanization, trade shocks and development programmes. Nevertheless, understanding existing capacity to adapt to present environmental changes represents a sound basis for starting to think about adaptation.

The second implication of the dynamism of island environments is that it complicates attribution of environmental changes to a single driver. Climate sceptics seize upon this dynamism to deny that environmental changes in islands are the product of anything other than natural processes, with some also attributing observed local changes to other (non-climatic) human activities. It is difficult, however, to say that climate change is not *also* a driver of environmental changes where local natural and human factors are clearly present, as the means for identifying the causes of change, and their relative contributions in complex and dynamic environments are not well refined (this is discussed further in Chapter 7). Uncertainty about drivers and their relative importance is a key reason why many scientists think natural science can help solve the problem of climate change in islands. If it were possible, clearer attribution of the drivers of change might be of significance for climate change negotiations. Nevertheless, uncertainty is not a reason for inaction. A precautionary approach to climate policy is called for in Article 3.3 of the United Nations Framework Convention on Climate Change (UNFCCC) which says that:

> *The Parties should take precautionary measures to anticipate, prevent or minimize the causes of climate change and mitigate its adverse effects. Where there are threats of serious or irreversible damage, lack of full scientific certainty should not be used as a reason for postponing such measures.*

The island environment as resource

The Pacific Islands considered in this book are all located within the tropics. However, they do not all share the same climatic characteristics. The climate of the region is dominated by the trade winds which blow air, laden with moisture evaporated from the warm ocean waters of the tropics, equator-ward and to the west. These winds meet at the Intertropical Convergence Zone (ITCZ), which migrates north and south of the equator according to the seasons. There is a second convergence zone, the South Pacific Convergence Zone (SPCZ), which lies to the south of the ITCZ. As a result of these climate patterns, the central western Pacific area tends to experience considerably more precipitation than the eastern Pacific, although the western islands are exposed to more intense droughts during El Niño events (see below) (Sturman and McGowan, 1999).

Many of the intra-plate and volcanic islands have both surface and groundwater systems, and because most lie within the central western Pacific they typically do not have the same level of exposure to drought. On high islands the main source of fresh water is orographic rainfall which gives rise to a distinctive pattern of wet (windward) and dry (leeward) sides of islands in the western Pacific. In terms of resources, this is typified by the wet, taro (*Colocasia esculenta*) / dry, yam (*Dioscorea spp.*) dichotomy in Pacific Island agriculture made famous by Barrau (1965).

Being low-lying, atolls do not cause orographic rain to fall and are dependent upon convectional rain. The rain water settles on the atoll surface, some evaporates and some percolates through the limestone to the water table. The fresh water sits upon the salt water which saturates the limestone substrate. Known as the Ghyben-Herzberg lens this freshwater resource is critical for the habitability of atolls and is prone to depletion and saline incursions during drought events and storms.

An important element of the climate of the region is the El Niño Southern Oscillation (ENSO) during which the usual low pressure in the western Pacific is replaced by high pressure, and lower pressures are measured over the central and eastern Pacific. During El Niño events, which occur with an irregular periodicity of two to ten years, the easterly trade wind pattern is disrupted: the easterly trades fall away and the western islands experience low rainfall; while the islands to the east, in the central part of the Pacific Ocean, experience greater than usual rainfall (Sturman and McGowan, 1999). In 1997 a particularly intense El Niño event led to drought and widespread food shortages in the Melanesian countries, and in particular Papua New Guinea (Bourke et al, 2001). In the highlands of Papua New Guinea an unusual but significant effect of El Niño is increased damage to crops from frosts. In these areas a cultural adaptation to frost has been the building of very large mounds upon which sweet potato are cultivated. The mounds enable the tubers to grow above the cold air which descends down the mound slopes (Waddell, 1975). During El Niño events the frosts become extreme, the mounds are unable to protect the plants and crop failure results (Allen, 1989, 1997).

Climate is not the only determinant of natural capital in the region. There are a range of terrestrial ecosystem types in the Pacific region that also provide the resource base upon which most communities still meet most of their needs (Manner et al, 1999). These range from strand and mangrove (both coastal forms) through savannah, and lowland tropical and montane rainforest. The biogeography of the region has a distinct west–east trend. Generally, Pacific Island fauna and flora originate in Asia and, to a lesser extent, Australia. Thus the continental type islands which are closest to the source areas tend to exhibit considerably greater biodiversity than the more isolated oceanic islands in the central and eastern Pacific. In addition, the plate-boundary islands are much larger and are characterized by a greater range of possible habitats. It is not surprising then, that these islands have a much greater depth of biological wealth than the smaller intra-plate or oceanic islands. The plate-boundary islands, given their western location, are also favoured by a wetter climate (El Niño events notwithstanding) and generally better developed soils. These islands also tend to have a much greater range of minerals than the oceanic islands.

In the most resource-constrained islands, such as the low-lying atolls, the supply of materials and food from terrestrial areas is limited to a few key tree species (e.g. breadfruit, coconuts and pandanus) and some basic crops such as swamp taro (*Cyrtosperma*). In these places the marine environment is a critical determinant of livelihoods (fisheries in particular are crucial to survival) and this is more the case than for other islands in the region. Important aspects of the marine environment include mangrove, lagoon and coral reef ecosystems which were key sources of protein for subsistence-based traditional, and continue to be for contemporary rural, communities. Beyond the reefs lie vast expanses of ocean which have considerable pelagic resources, in particular skipjack and yellow fin tuna (Adams et al, 1999). While the total land area of the Pacific Island region is around 550,000 square kilometres, the area of all of the countries' exclusive economic zones (EEZs) totals some 28.7 million square kilometres. Some countries with tiny land areas have vast EEZs. Kiribati, for example, has a total land area of 810 square kilometres but an EEZ of over 3.5 million square kilometres, earning access fees paid by deep water fishing nations to the value of about 20 per cent of GDP) (Barclay and Cartwright, 2007).

Table 2.3 summarizes the broad differences in the abundance of natural capital between islands in the east and west of the region. Table 2.4 shows per capita fish consumption in 12 countries, demonstrating that consumption is typically highest in the countries where terrestrial resources are most limited (such as Tuvalu, Kiribati, Niue). However, as we shall see, despite these differences in the availability of resources, some of the highest population densities are found in the smaller oceanic islands in the east.

Table 2.3 *Indicators of natural resource abundance between islands in the east and west of the Pacific Islands region*

Resource indicator	Western Pacific	Eastern Pacific
Island size	Large	Small
Topographical diversity	Large	Limited
Rainfall	High	Low
Biodiversity	High	Low
Minerals	Many	Few
Overall resource base	Large	Small

Table 2.4 *Annual per capita fish consumption in selected Pacific Island countries*

Country	Annual per capita fish consumption (kg)
Cook Islands	34.9
Federated States of Micronesia	69.3
Fiji	20.7
Kiribati	62.2
Nauru	55.8
Niue	79.3
Papua New Guinea	13
Samoa	87.4
Somolon Islands	33
Tonga	20.3
Tuvalu	110.7
Vanuatu	20.3

Source: Bell et al, 2009

The island environment as hazard

Pacific Islands are exposed to a large range of hazards, as indicated in Table 2.5. Given the role of plate tectonics in their formation it is not surprising that geological hazards such as volcanic eruptions and earthquakes are significant, especially for the continental type islands along the subduction zone between the oceanic and continental plates. Recent volcanic high islands are also subject to eruptions. Climatic hazards are also extremely important in the region, and tropical cyclones are perhaps among the most prominent of these. Cyclones are 'revolving' storms that form within 5° to 30° north and south of the equator, over warm ocean waters (and are therefore usually seasonal in occurrence), and which migrate in a direction generally away from the equator. Only a few parts of the Pacific Island region are not exposed to tropical cyclones including inland Papua New Guinea, and locations near the equator such as northern parts of Papua New Guinea, Nauru and Kiribati, although these places can be affected by storm surges created by nearby cyclonic events. Cyclones often cause great devastation, not just because of the very high wind speeds that are generated but also because of the heavy associated rainfalls and storm surges that are generated. Contemporary Pacific Island economies are sensitive to damage from cyclones, and the costs of recovery and the dependence on aid donors as a result of recovery processes can have lasting political and economic effects. Barker (2000) goes so far as to argue that in Niue cyclones have powerfully shaped the structure of contemporary society through their effects on out-migration and aid dependence.

Table 2.5 *Examples of hazards in Pacific Island countries*

Geophysical		Biological	
Climatic	Geological	Floral	Faunal
Drought	Earthquake	Fungal disease	Ciguatera
Flood	Erosion	– sigatoka disease	Dengue fever
Frost	Landslide	Infestations	Malaria
Heatwave	Tsunami	– weeds	HIV/AIDS
Tropical cyclone	Volcanic eruption		Infestations
High waves			– cane toads
			– rhinoceros beetle

Source: After Burton, Kates and White (1993)

Drought is also a relatively common hazard, although it is much less likely to be reported unless it is extremely severe. Contemporary food production systems in the region are sensitive to drought because crops in the region are now rarely irrigated. Drought is a particular problem in atolls because they have a fragile freshwater resource base which can quickly be depleted if rainfall drops. But droughts can also affect high islands and often cause serious losses in agricultural productivity. An important influence over climatic hazards in the Pacific region is the El Niño phenomenon. As noted earlier, the western Pacific typically becomes much drier than usual and harsh drought conditions (and severe frosts in the Papua New Guinea highlands) are experienced. In addition, tropical cyclones which are typically generated in the western Pacific become more common in the central part of the region during El Niño events.

The region is also exposed to a number of biological hazards, which may be affected by climate. Foremost among these are vector-borne diseases such as malaria and dengue fever. Malaria is endemic in Vanuatu, the Solomon Islands and Papua New Guinea, while episodes of dengue fever are experienced periodically in most parts of the region. Incidences of ciguatera fish poisoning, caused by biomagnification of marine toxins in reef fish, increase during periods of higher sea-surface temperatures (Hales et al, 1999). Variations in rainfall and temperature can also influence the incidence of water-related diseases such as diarrhoea, which can have implications for child health in particular (Singh et al, 2001). The Pacific region is also prone to some crop diseases such as taro blight, which devastated Samoa's export industry in the 1990s. A similar episode in Bougainville in the mid-1940s resulted in 3000 deaths from malnutrition (Brunt et al, 2001).

Early transformations: Traditional human environment systems

Humans settled in the islands of the Pacific region over a long period. It is likely that communities lived in Papua New Guinea at least 40,000 years ago, while the more isolated eastern islands were settled within the last two millennia (Kirch, 2000). It seems that early human inhabitants brought considerable environmental change to the various islands, introducing plants and animals and interacting with the original environments, and in some cases, causing considerable reduction in biodiversity (Kirch, 1997b). While hunting and gathering played an important role in most Pacific Island livelihoods, most communities also practised some form of agriculture, probably with introduced cultivars such as taro and yams. Swidden agriculture, or shifting cultivation, was important, particularly in tropical rainforest areas. While swidden agriculture was found on virtually all high islands and raised atolls with suitable forest conditions it was often linked with other agricultural practices such as arboriculture, irrigation (for taro) and drainage (yams), intensive farming systems with permanent fields, and animal husbandry (Kirch, 1997b; Clarke et al, 1999).

There is quite a degree of debate about the relationships between Pacific Island communities and their environments prior to the intrusion of beachcombers, sandalwood traders, whalers and missionaries in the late-18th and 19th centuries. On the one hand there is considerable support for a notion of pre-European island communities existing in some kind of ecological equilibrium with their insular environments. Perhaps the roots of such a view could be found

in the views of early European writers who conflated indigenous South Sea Islanders with nature. For example, Bougainville, one of the earliest Europeans to go to
Tahiti, was struck by its paradisiacal and fertile qualities. His travel companion,
Philibert Commerson, familiar with Rouseau's notion of *l'homme naturel*, wrote
popular accounts of the people of Tahiti living in a natural state (Spary and
White, 2004). Early cultural ecologists showed that traditional communities
maintained states of dynamic equilibrium with their island environment through
a range of rituals and other cultural practices (e.g. Rappaport, 1968). In an article
that was somewhat ahead of its time, William Clarke (1977) outlined what he
called 'structures of permanence' in Bomagai-Angoiang agro-ecosystems in high-
land Papua New Guinea. These systems were: 'palaeotechnic', not requiring
external subsidies of energy or nutrients as in modern agriculture; non-self poisoning; had strongly positive net energy yields; had a human-bound time scale
(not geological as in fossil fuel based systems); control of resources was equitable
and self-sufficient; resources were seen as capital to be preserved; and the agricultural system was based on a high level of species diversity. In today's language,
such structures of permanence may well be called sustainable agro-ecosystems.
Fisk (1962), described the conditions of the traditional agricultural systems of
Papua New Guinea, in some areas relatively undisturbed by the market economy,
as being in a state of 'subsistence affluence', and this is a notion others such as
Sahlins (1972) and Thaman (1994) have used in the context of Pacific Island
socio-ecological systems.

Others, however, dispute such a view, highlighting numerous examples of
archaeological evidence of widespread environmental degradation (deforestation
and erosion) that occurred following human colonization of the islands (Kirch,
1997a, 2000). In particular Rapanui (Easter Island) has been singled out as an
example of the negative outcomes arising from environmental mismanagement
(e.g. Flenley and Bahn, 2003; Diamond, 2005), though others claim that much
of the degradation may have occurred after European contact (Rainbird, 2002).
Not all environmental change that did occur in pre-colonial times was necessarily
bad: in some cases a kind of landesque capital was produced. For example, Spriggs
(1986) suggests that on Aneityum (in Vanuatu), the first arrivals cultivated the
relatively poor soils on the slopes causing deforestation and erosion, eventually
leading to the development of extensive alluvial flood plains where formerly
swamplands existed and which have since been settled and have formed the basis
for intensive agriculture. Yet others posit that it is not so much people that were
the principal drivers of changes in island environments in pre-colonial times, but
rather that it was climatic changes in the past 1250 years, such as the 'Little Ice
Age', that caused the most significant impacts on Pacific Island social-ecological
systems (e.g. Nunn, 2007). Whatever the rate of environmental degradation and
its causes, communities have been established and have survived on a large proportion of the islands in the region.

That such communities of Pacific Islanders have been sustained over time indicates that they can live sustainably. Some of the environments in which communities have survived are remarkably inhospitable, and that communities have survived in them for so long is testimony to their ability to adapt to difficult circumstances. Consider, for example, the communities that have survived on atolls for at least 2000 years. Many atolls are very isolated and all have limited or no soil, with the land composed almost entirely of sand. Their vegetation is typically strand and provides only a few plants that contribute to human livelihoods. With no natural soil and only the Ghyben-Herzberg lens, crop production is also very difficult and a very narrow range of cultivars is found on atolls. The key crop is the swamp taro which is cultivated in woven baskets placed in pits dug to reach the water table and filled with vegetable matter to form a compost-based anthropogenic soil. The key source of protein is fish, which is caught from within lagoons or in the deep ocean from canoes. Traditional canoe fishing is a hazardous activity and knowledge about where and how to catch fish was critical to the survival of atoll peoples – so much so that those who possess this knowledge still remain highly regarded within their communities (Hooper, 1983).

Traditional social-ecological systems also had to cope with extreme events, particularly drought and tropical cyclones. Throughout the region it would appear that there were numerous practices that enhanced the resilience of communities in the face of such events, as described by Lessa (1964), Marshall (1979) and Campbell (1990), among others. These practices ensured, whether purposefully or not, that communities would sustain an adequate nutrition supply after all but the most destructive events. Practices included surplus production, food storage and preservation, crop diversity, land fragmentation, use of famine foods and planting highly resilient crops. Systems of intercommunity cooperation were also important, often strengthened during times of plenty by ceremonial exchanges and underpinned by surplus production. In times of hardship these links could be called upon for assistance (Campbell, 2006). In several PICs, the susceptibility of houses to wind damage during tropical cyclones was reduced by settlement patterns which avoided cyclone-prone coastal areas, and methods of housing construction which were able to resist storm damage, or when damaged could be relatively easily rebuilt.

Whatever the theoretical position one might take, a great many Pacific Island communities appear to have been thriving at the time when Europeans first intruded into the island realm, where they found lively populations and described the environments as gardens of Eden (Withers, 1999). As Kirch (2000) points out, many Pacific Island communities did have viable agricultural systems and prudent systems of resource management, but not all were successful. Further changes to the human environment system were to come about in the period of colonization.

Colonization and environmental transformation

Prior to the 19th century there had been very little colonial activity in the Pacific region, with only Spain establishing sovereignty over Guam and the remaining Mariana Islands. By the middle of the 19th century through to the early 20th century France, Germany, the United States and the United Kingdom began to expand their empires into the region. Japan, Australia and New Zealand took over German colonies during the First World War, and the United States took over Japan's control of most of Micronesia (with the exception of Kiribati and Nauru which were British colonies) after the Second World War. Niue and the Cook Islands were also transferred from the United Kingdom to New Zealand in the early part of the 20th century (Kiste, 1994).

While slow changes had taken place in the social-ecological systems of the Pacific Islands in the centuries that passed between initial settlement and the arrival of Europeans, colonization and the introduction of capitalism, Christianity and its attending values was a far more dramatic driver of change thereafter. However, initially the daily routines of most Pacific Island communities experienced relatively little change (Campbell, 1989).

Perhaps the first major impact on social-ecological systems was the introduction of diseases to which Pacific people had no immunity. In a number of Pacific Islands 'blackbirding', the taking of indentured labourers to Queensland, Fiji and even South America, also contributed to population decline, as a large proportion (about a quarter) did not return (MacArthur and Yaxley, 1968). While the levels of depopulation were variable, in some islands pre-colonial populations were reduced by more than 90 per cent (Spriggs, 1997). Population loss, beginning mostly in the 19th century, continued into the mid-20th century, with many island populations reaching their nadir around the time of the Second World War. Since then populations have grown and now population growth and high population densities are bringing about environmental problems, particularly in rapidly growing urban areas (Cocklin and Keen, 2002).

The initial decline in human numbers had environmental implications. The demand for land dropped considerably and pressure on resources for subsistence purposes may also have reduced. On the other hand, there were new demands placed on the resource base arising from the introduction of taxes, church fees and the cash economy, as well as the introduction of commercial crops such as copra (the dried meat of the coconut). The decline in demand for food production enabled coconut plantations to be expanded in many places (Campbell, 1990). Meanwhile land was also alienated for foreign-owned plantation development. Effectively, large areas were removed from subsistence production.

Another outcome of colonization was increased vulnerability to extreme events. A number of social and economic changes reduced the resilience of Pacific Island communities in the face of natural hazards. This is illustrated in Figure 2.2, which shows that there has been both a reduction in community resilience and greater exposure to extreme events than before colonization. The cash economy enabled people to purchase rice and other foods and reduced the need for food preservation and storage, the traditional social and religious practices which underpinned many ceremonial exchange systems were proscribed by colonial governments and missionaries, and newly introduced crops such as cassava (*manihot esculenta*) replaced the sturdier, more wind resistant, traditional cultivars such as yams and taro (Campbell, 2006). Many communities went from surplus food producers to deficits because coconut plantations, and later a variety of commercial crops, expanded onto food-producing land. When populations began to recover to their pre-colonial levels in the second half of the 20th century, there was considerable pressure on the lands still available for food production. In many places, cassava became a mono-cropping staple because it had lower requirements of land fertility and labour. Thus there was a reduction in agro-diversity, expansion of introduced and more vulnerable (to wind damage) crops, and decreasing per capita land availability for food production, all of which reduced food security in much of the region.

The increased vulnerability to extreme events brought about by colonization was partially offset by the provision of disaster relief by colonial governments, which tended to be relatively ad hoc until the second part of the 20th century when it became an expected response. In the longer term, disaster relief has created something of a moral hazard in many places, further undermining resilience by creating dependency on external resources for disaster planning, response and recovery.

At the same time as colonization rendered communities in the Pacific more susceptible to damage from disasters, it also increased their exposure to extreme events. Whereas many traditional Pacific Island communities lived in small hamlets and were often located on high land for defence purposes, the colonial authorities, in cooperation with missionaries, successfully encouraged amalgamation and the establishment of coastal villages (Spriggs, 1997). This placed many communities at increased risk from tropical cyclone events through increased exposure to storm surge, with further implications for exposure to climate change and sea-level rise. Colonialism also set in motion the process of urbanization with colonial administrative centres being the first urban places to emerge in the region. Significantly, the great majority of these urban places are located on the coast and likely to be exposed to the coastal hazards associated with global warming. These changes, coupled with growing populations, increased development of tourism facilities (usually in coastal locations) and growing accumulation of material possessions all increase the material and human losses in the event of extreme events. This is of considerable significance given that it is likely that the frequency and/or intensity of climatic extremes may increase as a result of global warming.

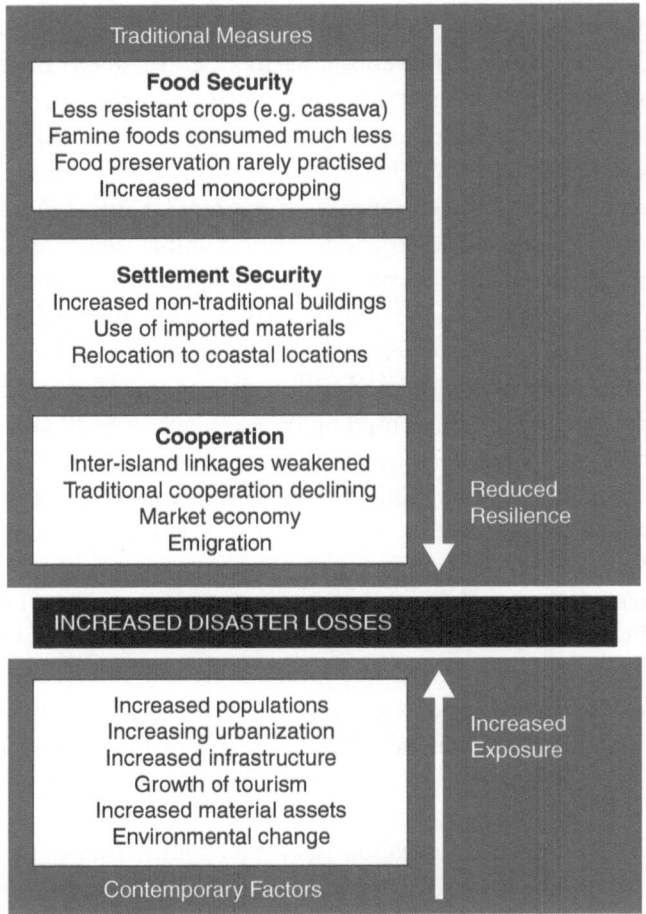

Source: After Campbell, 2009

Figure 2.2 *Factors undermining resilience and increasing exposure to hazards in Pacific Island communities*

The post-colonial Pacific: Environment and development

Thus far we have described the exposure of Pacific Island social and ecological systems to environmental changes, and we have touched on their sensitivity to these changes. This discussion has also taken account of the history of human environment relationships in the region. We turn now to consider the contemporary Pacific, and many of the social factors that influence the generic capacity of Pacific Island societies to adapt to climate change.

The first Pacific Island country to become politically independent was Samoa in 1962. The next two decades or so saw a further 13 countries achieve similar status (including the Cook Islands and Niue, which became self governing in free association with New Zealand, and the Republic of the Marshall Islands, the Federated States of Micronesia and Palau, which are constitutional governments in free association with the United States) (Henderson, 2002). However, economic independence has not followed political independence: the economies of PICs have few exports of significant value, are almost entirely dependent on imports for manufactured goods and technology, typically have high trade deficits and rely heavily on aid. As explained later, this dependence on external forces does not necessarily equate to mendicancy.

The last two decades of the 20th century, and the beginning of the 21st, have seen a number of new trends emerge that are slowly changing some aspects of life in the Pacific. Communities have become increasingly linked to the global economy, many Pacific Island governments have adopted neo-liberal economic policies, tourism has been an expanding sector in many countries, urbanization – as we have seen – has increased rapidly, and a number (but certainly not all) of countries have developed large (sometimes exceeding national populations) diasporas in New Zealand, Australia and the western United States. Table 2.6 provides basic economic data for the PICs as well as the Human Development Index for each country.

Development issues in post-independence PICs

Colonization initiated a process that has seen PICs become increasingly integrated into the global economic system, although it is still the case that global economic forces only partially (if at all) shape social life in many places in the region. The towns that formed during the colonial period provided the loci of urbanization, and now for the most part are the nodes through which imports and exports enter and leave the PICs. Over time the dependence of island economies on imports has grown, such that only Papua New Guinea has a positive balance of trade (reflecting its mineral wealth) (see Table 2.6). In most cases trade imbalances arise because of the costs of imports of petroleum products, with food also being a significant component of imports in many countries. Urban areas have been and continue to be the main point of departure for migrants, who travel to Pacific rim destinations, especially New Zealand, Australia and the United States (west coast) where many Pacific Island societies, especially those in Polynesia, have very large expatriate populations (Bedford, 2000). In extreme cases, such as the Cook Islands and Niue, where the people are New Zealand citizens, there are many more people of Cook Island and Niuean descent living in New Zealand than there are in living in the islands. Indeed, even the majority of people born in these countries live in New Zealand: In Niue's case approximately 80 per cent of all Niueans born in Niue now live in New Zealand (Bedford et al, 2006).

While colonial governments partly provided health and education services, and urban infrastructure, there was less commitment to other forms of 'development' such as rural transport and energy systems, value-adding forms of production, or building the capacity for self-government. Thus throughout the

Table 2.6 *Development characteristics of Pacific Island countries*

Country or Territory	Total GDP ('000 USD) (2006)	GDP per capita (USD) (2006)	Annual trade balance ('000 USD)	Consumer Price Index (2007)	HDI* (2005)
MELANESIA					
Fiji Islands	2,695,666	3,175	-1,058,563 (2007)	4.8%	0.762
New Caledonia	7,129,631	29,898	-720,120 (2007)	1.8%	n/a
Papua New Guinea	6,044,220	991	2,285,892 (2007)	3.2%	0.530
Solomon Islands	373,800	753	-101,474 (2006)	7.6% (2006)	0.602
Vanuatu	459,010	2,127	-175,400 (2007)	3.9%	0.674
MICRONESIA					
Fed. States of Micronesia	235,900 (2007)	2,183 (2007)	-134,572 (2006)	3.5%	0.569 (1998)
Guam	3,700,000 (2005)	22,661 (2005)	-596,177 (2007)	6.8%	n/a
Kiribati	61,433	653	-56,997 (2006)	-1.6% (2006)	0.515 (1999)
Marshall Islands	149,219 (2007)	2,851 (2007)	n/a	4.3% (2006)	0.563 (1998)
Nauru	27,661	2,807	-21,830 (2005)	n/a	0.663 (1998)
Northern Mariana Islands	948,659 (2005)	12,638 (2005)	n/a	2.1%	n/a
Palau	170,144 (2007)	8,423 (2007)	n/a	6.6%	0.861 (1998)
POLYNESIA					
American Samoa	437,900 (2003)	6,995 (2003)	-140,700 (2006)	3.0% (2006)	n/a
Cook Islands	182,175	8,553	-170,328 (2007)	2.4%	0.822 (1998)
French Polynesia	5,640,452 (2004)	22,474 (2004)	-1,601,301 (2007)	1.9%	n/a
Niue	10,006 (2003)	5,828 (2003)	-7,536 (2004)	6.7% (1998)	0.774 (1998)
Samoa	532,000 (2007)	2,872 (2007)	-167,356 (2007)	6.1%	0.785
Tonga	234,484	2,319	-106,170 (2006)	5.8%	0.819
Tuvalu	17,514 (2002)	1,831 (2002)	-12,698 (2006)	2.2%	0.583 (1998)

* Human Development Index, a measure of the average achievements of a country in terms of a long and healthy life, knowledge, and standard of living

Source: SPC, 2008a; UNDP, 1999; UNDP, 2007

colonial period, most rural communities engaged in a dual economy of mixed subsistence and cash primary production, the ratio of which reflected proximity to urban centres as much as anything else. Outer island communities typically earned minimal cash incomes, and what was earned was largely through copra production and (circular) labour migration. In the main, rural communities were self-sufficient, relying on subsistence production through agriculture, consumption of wild plants and animals, and the use of marine resources. The colonial era has thus been described as one of continuity through change (Brookfield with Hart, 1971).

An important question that has arisen in the post-colonial era is the extent to which this continuity has been maintained. Given that urban settlements did not exist in pre-contact Pacific Islands (Connell and Lea, 2002), and remained relatively small in the colonial era, the growth of urbanization since independence may be seen as a significant discontinuity. Table 2.7 shows the extent of urbanization in each of the PICs. Clearly, there is considerable variation, but some countries have large urban populations, and some have urban areas that are growing very rapidly. While urbanization has had important environmental implications, which we discuss below, many so-called Pacific Island 'traditions' are still sustained in urban areas – for example kinship is still important, and taro, yam and fish are still key staples. Nevertheless, urban residents in the region have more constrained access to the means of subsistence production (though as Thaman (1995) observes, urban agriculture is still important), and are more dependent upon salaries or wages for their livelihoods.

For many urban dwellers, under-employment, unemployment and, increasingly, poverty are elements of their lives. Squatter settlements have emerged in many Pacific Island towns and cities where much of the land is held in communal ownership of the original clans or families. The buildings in these areas often lack the qualities of traditional dwellings that are resistant to tropical cyclones (Campbell, 2006). People living in these peri-urban settlements often have poor access to sanitation infrastructure and clean water, and so typically have higher rates of water-borne diseases. The quality of infrastructure in these areas is constrained by uncertain land tenure: households and public authorities are reluctant to invest in areas liable to be reclaimed by customary owners. Because urban growth has failed to successfully adjust to land tenure arrangements across the region, but also because of the costs of urban development and the lack of capacity in urban design and management, urban areas in PICs have a chequered history with respect to effective urban planning. Accordingly, the infrastructure and people in urban areas in PICs are vulnerable to natural hazards. Given this, there are many challenges – some of which may be intractable – associated with implementing measures to ensure that urban areas and people in the Pacific are able to adapt to climate change.

Table 2.7 *Urbanization in Pacific Island countries*

Country or Territory	% of Population in urban areas	Total urban population (2008)
MELANESIA		
Fiji Islands	51	428,000
New Caledonia	63	155,400
Papua New Guinea	13	841,600
Solomon Islands	16	82,800
Vanuatu	21	48,900
MICRONESIA		
Fed. States of Micronesia	22	24,300
Guam	93	166,500
Kiribati	44	42,800
Marshall Islands	68	36,200
Nauru	100	10,200
Northern Mariana Islands	90	56,700
Palau	64	13,000
POLYNESIA		
American Samoa	50	33,100
Cook Islands	72	11,200
French Polynesia	53	139,500
Niue	36	600
Samoa	21	37,800
Tonga	23	23,600
Tuvalu	47	4,600

Source: SPC, 2008a

PICs face a variety of constraints to their economic development. Isolated economies struggle to compete in global markets due to the costs of sustaining reliable transport routes over large distances (Briguglio, 1995). In all but the large Melanesian countries there are insufficient factor endowments and inadequate levels of domestic demand for the emergence of economies of a scale necessary to compete in global markets. In almost all countries labour costs are relatively high compared to the major Asian economies, yet labour is relatively unskilled. The small infrastructure base also creates inefficiencies in the movement of goods and services.

Thus primary products dominate exports in the Pacific, with the main goods being minerals (from the Melanasian countries), timber (from Papua New Guinea and the Solomon Islands) and fish (largely in the form of licenses paid to countries by foreign fishing fleets). A few countries have significant mineral reserves, most notably Papua New Guinea, which has several large gold and copper mines, many of which have had deleterious social and environmental effects (Böge, 1999; Banks, 2002; Walton and Barnett, 2008). More recently reserves of natural gas and oil have begun to be exploited in Papua New Guinea. New Caledonia produces nickel, Fiji produces gold, and a number of raised limestone islands such as Nauru, Banaba (in Kiribati), Makatea (in French Polynesia) and Angaur (in Palau) have been mined for phosphate. Commercial logging has been restricted mostly to Melanesia, and there has been considerable deforestation in Papua New Guinea and Solomon Islands in particular. Most countries lack the financial and technical capacity to take full advantage of the fisheries resources in their large and abundant EEZs. Value adding to primary products remains minimal, restricted to small- to medium-scale enterprises producing timber veneer, canned tuna, sugar, and niche commodities such as handcrafts, coconut based soaps and skin crèmes, organic vanilla extract, noni juice and fresh chilled fish. FIJI Water has become a popular bottled water, in a burgeoning global market, a case in which the image of Fiji as a south seas paradise, natural and exotic is marketed as much as the water inside the bottle (Connell, 2006). Other Pacific countries are attempting to follow suit.

The manufacturing sector of most PICs is small or non-existent, with the garment industry in Fiji and the Northern Mariana Islands, and an automobile parts plant in Samoa being notable exceptions. The larger towns have a number of service industries oriented largely towards domestic and regional demand. The viability of many of these is under threat from emerging trade liberalization agreements which may lead to increased competition from international firms (Narsey, 2004).

Tertiary sector activities are dominated by tourism, which is a large source of foreign revenue in countries such as the Cook Islands, Fiji, French Polynesia, Guam, the Northern Mariana Islands and Palau. Much of the tourism is resort oriented, but in some countries with established markets (such as Fiji) 'village stay' type tourism and small-scale ecotourism are showing some growth, though numbers of such travellers are still relatively small. There are some specialized forms of tourism also developing, such as diving among World War II wrecks in the Marshall Islands, and observing cultural rituals such as land-diving on Pentacost Island in Vanuatu. There is some trepidation in the tourism industry about climate change. Aviation is a major contributor to greenhouse gas emissions and moves to mitigate greenhouse gas emissions could impinge upon the tourism sector if air travel were curtailed. Warmer climates in the major countries from which tourists to the region come from may also reduce demand, as may increased health risks in destination countries. The impact of increasingly intense climatic extremes on tourism infrastructure may also reduce the ability of tourist economies to supply accommodation and other services (Becken and Hay, 2007).

Some of the smallest countries have used novel approaches to raising revenue, based on the commercialization of their sovereignty (Palan, 2003). This includes strategies such as producing stamps for the philately market, and trading aid donors against each other in exchange for access to strategic islands or votes in the United Nations system (Poirine, 1999). Other examples include the renting of internet domain names: Tuvalu earns revenue from users of its '.tv' internet domain name, and Niue ('.nu') and Tonga ('.to') have also received some rent for their domain names (Wilson, 2001). Some countries have at times earned foreign exchange from granting passports to foreigners, and from hosting offshore finance centres of various kinds: both practices have in the past been significant sources of revenue, but were curtailed in the wake of the September 11 attacks in New York when such schemes were seen as having the potential to be misused by terrorists (Hampton and Christensen, 2002; Eden and Kudrie, 2005).

The particular characteristics of the economies of the small PICs led Bertram and Watters (1984, 1985) to describe them as MIRAB economies, so called because of the significance of migration, remittances, aid and bureaucracy to revenue. The claim is not without empirical basis: then as now these activities, which are not based on production but on transfer payments and non-tradable production, dominate the economies of many PICs. For example, remittances are equal to approximately 15 per cent of GDP in Kiribati and Tuvalu, 25 per cent in Samoa, and close to 40 per cent in the case of Tonga (Boland and Dollery, 2007; Browne and Mineshima, 2007). Across the region, remittance flows have grown by over 30 per cent in the past decade, and the prospects for future growth are good (Browne and Mineshima, 2007). The Pacific is nevertheless relatively unique in that aid flows are still larger than remittances. Relative to GDP, some countries in the Pacific are the biggest recipients of aid in the world, with aid flows equal to 27 per cent of GDP in Palau, 34 per cent of GDP in Kiribati, 51 per cent of GDP in Federated States of Micronesia, 61 per cent of GDP in the Marshall Islands, and possibly as much as 80 per cent of GDP in Niue (Sampson, 2005; Barnett, 2008a). These aid transfers sustain large bureaucracies. For example, government spending is equal to more than half of GDP in The Federated States of Micronesia, Kiribati, the Marshall Islands, Niue, Palau, The Solomon Islands and Tuvalu (Commonwealth of Australia, 2008).

There is some debate about the validity and usefulness of the MIRAB description of small Pacific Island economies (James, 1993; Fraenkel, 2006). Part of the debate has been about the extent to which MIRAB, and associated notions of rents, frames the region's economies and peoples in pejorative terms – as 'rent seekers' who have no reason to be productive members of the modern economy. Yet there is little hint of this in Bertram and Watter's early or subsequent work, which, along with others' analyses show that a dependence on non-productive flows does not makes Pacific Island leaders mendicant – indeed they display remarkable agility in managing to sustain transfer payments over time (Bertram, 1999). Migration and the provision of remittances is a rational response to development aspirations in Pacific communities, and it is important for members of Pacific Island diasporas to maintain their links with their kin in the Pacific (Brown, 1997; Hooper, 2000). It should also be noted that bilateral donors obtain benefits from their aid relationships (Sevele, 1987). From these perspectives, MIRAB economies arise from fair exchanges, even if those exchanges are not necessarily ones that are based on the export of goods.

Most PICs have a longstanding commitment to sustainable development and, recognizing the importance of the environment to their economies and cultures, they have actively participated in global and regional meetings and agreements on environmental issues from their beginnings in the early 1970s. They have also been actively involved in both the Barbados and Mauritius meetings on sustainable development in small island states (Fry, 2005). In addition they have established a regional set of millennium development goals (SPC (Secretariat of the Pacific Community), 2004) and recently developed the Pacific Plan, which seeks to promote sustainable development in the region through regional cooperation, pooling resources and services and integrating the economies of PICs (PIFS (Pacific Islands Forum Secretariat), 2007). Yet despite these past and present efforts, many of the kinds of political and economic activities described in this chapter have caused and continue to cause environmental problems, which in turn increase the sensitivity of social-ecological systems to the potential impacts of climate change.

Environmental issues

While PICs do not share the high levels of environmental degradation found in many of the developing countries where heavy industry is a path to growth, there remain a number of serious environmental problems apart from the effects of climate change. These are summarized in Table 2.8. The causes of and solutions to these environmental problems must be understood in the context of low levels of economic and human development in the region, and subsequent state and community expectations for better standards of living. Environmental problems in the region are intricately linked to economic development and thus they have tended to increase in scale and intensity as the imperatives created by markets have come increasingly to influence social-ecological relationships. Most export markets are for primary resources extracted at minimal cost, and with minimal value-added, which constrains the generation of wealth necessary to achieve the kinds of improvements in some environmental problems that are observed elsewhere in the developing world (see Torras and Boyce, 1998). Markets have not emerged to deal with environmental problems, and low levels of growth limit the resources available to the state to manage environmental problems.

Table 2.8 *Environmental problems in Pacific Island countries (excluding climate change)*

Problem	Proximate causes	Direct effects
Waste management	Increased waste production per capita	Land pollution
	Increased solid and liquid waste generation	Fresh water pollution
	Inadequate disposal facilities	Coastal pollution
	Increasing use of agricultural chemicals	
	Stockpiles of persistent organic pollutants	

Table 2.8 (continued)

Problem	Proximate causes	Direct effects
Land degradation	Deforestation leading to erosion Increased agricultural intensity Expansion of agricultural land Logging	Silting (rivers, lagoons and reefs) Reduced soil fertility Increased river flooding
Coastal degradation	Overfishing and gathering Pollution and silting Removal of mangroves Damage to coral Sand mining Coastal development including tourism	Loss of biodiversity Increased coastal erosion Loss of habitat Loss of resources
Freshwater degradation	Watershed degradation (e.g. deforestation) Over use of ground water Increased demand (especially urban) Waste management High levels of wastage from supply systems	Reduced freshwater quantities Reduced freshwater quality
Reduction of biodiversity	Deforestation Agricultural expansion Wetland and mangrove clearing River, lagoon and reef sedimentation Urbanization Pollution Invasive species	Loss of valuable species – ecologically – economically – culturally

Thus, many of the environmental problems that were created in the colonial period have remained, and some, such as deforestation, have worsened in intensity as communities seek to engage in both subsistence and commercial agriculture by expanding onto previously unfarmed slopes, or intensification of agriculture in areas formerly used for swidden production, or have sought royalties from logging companies. New problems have also emerged with the growth of urban areas where water supply, waste management and pollution are significant issues.

Waste management is particularly problematic for the smallest island countries, where the costs of effective systems for disposing of solid and liquid wastes are beyond the means of governments, and where small land areas mean even small volumes of waste cause environmental, health, and aesthetic problems. In addition, exporting waste, even for recycling or re-use, is not economically feasible given high transport costs. As a result, the non-biodegradable wastes arising from consumption of imported products (very little non-biodegradable waste is produced endogenously in PICs), including plastic bags and bottles, nappies and oils accumulate in island ecosystems.

In Tuvalu, for example, even the small amount of waste produced by the 4600 people living in Funafuti creates (an eyesore and) environmental problems such as contamination of the freshwater lens (see Figure 2.3). Upon witnessing this waste, it is tempting to conclude that Tuvalu's environmental problems, including its vulnerability to climate change, are of its own doing. Connell (2003), for example, considers it ironic that while Tuvalu actively campaigns on climate change in international forums, it fails to address unsustainable local environmental practices. This somewhat ignores that Tuvalu is a Least Developed Country, where per capita consumption of imported products is small relative to people living in most developing and all developed countries, and that there is no safe or effective place to dispose of non-biodegradable waste in an atoll environment. Tuvaluans have a right to and want some of the benefits of modernity, such as electricity, a diverse diet, western medicine, telephones, the internet and consumer goods such as radios and televisions. However, their environment struggles to assimilate even the smallest of wastes arising from this consumption; and it remains the case that if Tuvaluans consumed and polluted to the same degree as did people in Europe, North America or Australia, they would be completely buried in waste. It is also the case that there are many features of social-ecological relations in Funafuti that very positively contribute to sustainability, such as: the

Figure 2.3 *Waste dump, Fongafale, Funafuti, Tuvalu*

widespread use of rainwater harvesting (see Figure 2.4); and the fact that while approximately three quarters of the population of Funafuti are from outer islands, only 7 per cent of households are living in impermanent shelters without access to a toilet or water on the premises – indicating a strength of kinship ties and a generosity and flexibility with respect to housing and land tenure that few if any societies in the world enjoy (Government of Tuvalu, 2005).

While Table 2.8 indicates the proximate causes of the environmental problems in PICs, it is important to identify the underlying causes. As we have seen, Pacific Island communities have experienced considerable change in the past century and a half. Incorporation into the global economic system has seen changes in land use that have resulted in the expansion of farming into areas that were formerly under forest, and fallow periods have declined in many rural areas with associated reductions in soil fertility. Logging and mining have added to this degradation, the former in particular, often illegal and associated with corruption (Barnett, 1990; Larmour, 1997). Unfortunately, in addition to the environmental costs, corruption often results in little benefit finding its way towards meeting the countries' development aspirations (Miranda et al, 2003).

Population growth is often considered to be a major factor in environmental degradation (McIntyre, 2005; Chape, 2006), although it is not altogether clear that countries in the region with high rates of population growth have greater levels of environmental degradation than those with low or even declining rates. Nevertheless, the regional annual population growth rate is estimated to be 1.9

Figure 2.4 *Rainwater tank, Fongafale, Funafuti, Tuvalu*

per cent (SPC, 2008b). This represents a doubling time of 37 years. Some countries have rapid rates of population increase, although three of the four highest, Papua New Guinea, Solomon Islands and Vanuatu, have the lowest crude population densities. On the other hand, the atoll countries have the highest population densities.

However, data about population density do not necessarily say much about environmental impacts. Two important factors in determining the environmental impacts of population in the region, as elsewhere, are: the technologies people use for modifying the environment (including modifications to manage environmental problems); and the way markets encourage environmentally damaging activities and the policy and measures available to limit these. Another important factor to consider when assessing population and the environment in any given place in the region is land tenure governance, which determines access to, and the use of, land and marine resources. Localized environmental impacts are most often more accurately described as situations where customary land tenure no longer exists or has been circumscribed in some way.

A more significant demographic trend than simple population growth is the increase in the number and concentration of people in urban areas. It is in urban areas that problems of pollution from solid and liquid wastes are most pressing, most energy is consumed, local air quality can be poor, and exploitation of land and marine resources is most intense. Many urban areas in the region do not have sanitary landfill sites, and waste from dumps, point sources, and poor or non-existent sewerage systems finds its way into soils and coastal ecosystems.

There are also some environmental problems in the region that local people cannot be said to have had any decisions over, such as the intense environmental and social impacts of phosphate mining, and the ecological and health effects of British, French and US nuclear weapons tests. In the Marshall Islands for example, the United States' nuclear weapons tests have rendered whole islands uninhabitable, and have affected large-scale forced migration within the country (Kuletz, 2001). Similarly, the vast majority of the land areas of Nauru and Banaba have been devastated by phosphate mining controlled by German, Australian, British and New Zealand interests (Gowdy and McDaniel, 1999; Teaiwa, 2005).

Institutional arrangements for environmental management are constrained in many PICs. The costs of running an environmental ministry or department, engaging in global environmental decision making, developing legislation and other mechanisms for local environmental management and then implementing and policing them are very high relative to the sizes of the economies. In addition there is a shortage of human capacity in, and limitations on the availability of necessary information for, environmental management (McIntyre, 2005). Moreover, many countries are eager for investment and often the possible environmental effects of some development projects are overlooked. In most countries incorporating environmentally sustainable development into overall development planning is progressing slowly. Attempts to 'mainstream' climate change adaptation and disaster risk reduction have, thus far, had limited success.

The kinds of environmental and social problems we have described in this chapter increase the vulnerability of Pacific Island social-ecological systems to climate change. Degraded ecosystems are less able to absorb changes and maintain structure and function; coral bleaching is most intense in areas where reefs are already stressed. Social systems under stress are also less able to avoid and recover from changes; it is the urban poor living away from their customary lands and in hazard-exposed areas that are often the most affected by storms. In this sense, adaptation to climate change is about practices to bring about sustainability, such as implementing effective systems for sanitation and waste disposal, improving access to health care, the adaptive management of resources, and small-scale income generation schemes for the most vulnerable communities (Barnett, 2001). The crux of the adaptation problem in many small island states is that they will have to achieve a degree of sustainability probably unprecedented in any modern state, in a biophysical and economic environment which in many respects allows little room for error, and in a very short space of time given the imminence of climate change.

Conclusions

From where I stand, I do not see the lost people of the South Seas, the defeated and the despairing, shrunken shadows of those who went before. What I observe are the proud descendants of some of the most remarkable explorers and settlers who ever lived. We carry the cultural and historical inheritance of ocean navigators of peerless skill and their courageous kin who crossed vast distances before the tribes of Europe had ventured forth from their small part of the earth. Our forebears populated islands scattered over the world's greatest stretch of water, covering a fifth of the planet's surface. It was one of the most amazing migrations in history, a triumphant testimony to human endurance, fortitude and achievement.

Ratu Sir Kamisese Mara, July 1999

In this chapter we have sought to portray the environmental and social characteristics of PICs in order to better contextualize their vulnerability to climate change. We have sought to show that they exhibit very high levels of social, cultural and environmental diversity, are both resilient and vulnerable, have been subject to, and in many cases coped very well with considerable change, and are likely to raise substantially different sets of issues in relation to climate change from other settings.

It is widely considered that the Pacific Islands are highly at risk to the kinds of changes likely to arise from climate change, principally because they are located in the tropics and are exposed to the vagaries of weather and climate in that zone, and because of their high ratios of coastline to land area. It is also argued that the social-ecological systems of the Pacific Islands are highly sensitive to the likely effects of climate change because many ecosystems and organisms are sensitive to changes in air and sea temperatures, and many subsistence and market-based activities are dependent on these climate-sensitive natural systems. On this basis the prospects for a prosperous Pacific appear bleak.

But this is a perhaps overly pessimistic position on the region and its future, one that emerges when examining the region in generic terms, and over time, in the way that we have in this chapter, and which many others also do. There are two reasons why such pessimism must not be overly indulged. First, we have found most of the specific communities in the PICs that we have visited to be robust and resilient in the face a raft of changes that have been imposed on them since colonization, and whose members go about their lives with dignity and, particularly in rural areas, obtain adequate mixed subsistence and cash livelihoods for their households. While the idea of Pacific Islands as gardens of Eden may have been dispelled, that tourists continue to be drawn to them indicates that they still have much to offer, particularly in comparison to the heavily industrialized and urbanized parts of the world that are ostensibly more 'developed'.

The second reason why there is room for optimism about the future is that the underlying framing of the Pacific Islands, which shapes all subsequent discussions of their vulnerability to climate change, in many respects puts too much weight on some aspects of island social-ecological systems, and not enough on others. The Pacific Islands are seen to be vulnerable to climate change because many of the determinants of generic adaptive capacity are seen to be lacking. Yet these determinants reflect a standard Western and continental idea of 'development' where strong economies, technological capabilities, good governance, good infrastructure, education and sustainable resource use are all seen to be highly desirable (Barnett, 2001). In many respects capacity to adapt is restricted by these factors, but it is also important to remember that the islands of the Pacific are not continents or parts thereof, and most are never likely to become modern export-oriented economies.

A more island-centric framing of the region suggests an alternative way to think about adaptive capacity. Epeli Hau'ofa outlines such an alternative framing, captured in the label 'Oceania', which he says 'denotes a sea of islands' rather than the prevailing Western view of 'islands in a far sea' (Hau'ofa, 1994, pp152–153). According to Hau'ofa, this is how grassroots Pacific people have historically viewed their world – not in terms of small, distinct and remote land areas as viewed by development economists, geographers and 'macropoliticians', but rather in terms of the expansiveness of the ocean, to be traversed for trade, marriage, adventure and knowledge. Hau'ofa argues that the geographic imagery of smallness and isolation was a colonial construct reified by cartographic delineation and perpetuated by its facilitation of imperialism.

That 'the world of Oceania is neither tiny nor deficient in resources' necessitates an alternative approach to understanding capacity to adapt to climate change in the Pacific (Hau'ofa, 1994, p156). As Hau'ofa and other anthropologists suggest, it is kinship and reciprocity that bind island communities together across space and time, and these are underweighted elements of adaptive capacity in many Pacific Island communities. For example it is these ties that sustain remittances (Brown, 1997) such that they continue to grow (Commonwealth of Australia, 2008). Remittances enable resilience to climate extremes as well. In response to the damage caused by cyclone Ofa in 1990 substantial sums of money and goods were remitted to Samoa by Samoans living in New Zealand (Campbell, 1998). Another implication of Hau'ofa's thesis is that adaptive capacity is more a matter of what islanders do in their oceans and coasts than what transpires on the land. This too is borne out in our discussion of food security, which is as much about fisheries and their management as it is about agricultural production.

If we adopt the indigenous view of Oceania as a 'sea of islands', rather than small islands in a large sea, then our understanding of the capacity of the islands to adapt to climate change requires rethinking. It demands a serious look at ways to enhance the benefits of migration and remittances to islands. It means considering and nurturing the value of kinship and traditional knowledge. It suggests that the assistance that comes from metropolitan countries focuses more on human development in ways that do not result in major ecosystem disturbances, and on the facilitation of *capacity* more than resource-based economic development. It demands an alternative approach to development that supplements the inherent skills and knowledge of Pacific people with appropriate technological and policy innovations. Ultimately it means creating the space and opportunities for Pacific people to do what they have always done with great success – adapt to change.

3

The History and Architecture of Climate Science

The Pacific Islands are home to some 9,500,000 people living on more than 10,000 individual islands. This makes broad claims to knowledge about the region epistemologically tenuous. Many of the things that are known about places or phenomena are often uncertain, and are a questionable basis for making generalized statements about the region. Moreover, what is formally 'known' is often highly fragmented and resides within, and rarely beyond, the boundaries of particular knowledge systems (such as academic disciplines).

Knowledge about climate change in the region, and in particular about the vulnerability of environmental and social systems and options to adapt to climate change, shares these characteristics of uncertainty, fragmentation and abstraction. This and the following chapter review how knowledge about climate change is produced and the way it may affect the region. This chapter begins with a survey of the development of scientific interest in climate change through key global research programmes, and examines the extent to which these large endeavours include the Pacific Islands. It then focuses on the Intergovernmental Panel on Climate Change (IPCC) process. The chapter ends with a discussion of the limits of these global research efforts, and the approaches they use. In so doing it describes the larger global research context that has shaped the nature of the regional climate change research initiatives discussed in Chapter 4.

Initial stirrings

The idea that the atmosphere traps solar radiation, and that changes in the composition of gases in the atmosphere could cause changes in climate, can be traced at least as far back as to the work of John Tyndall in the 1860s. Contemporary concern for the problem of climate change began when concentrations of carbon dioxide in the atmosphere were observed to be increasing in the early 1960s. This stimulated the development of research networks investigating the problem, such as the Study of Man's Impact on Climate (SMIC) group, which reported to the

1972 United Nations Conference on the Human Environment (the Stockholm conference) (SMIC, 1971). From the mid 1970s onwards the principal tools used to improve understanding of the relationship between greenhouse gases and climate were atmospheric general circulation models (GCMs). GCMs developed at the Lawrence Livermore National Laboratories, the National Center for Atmospheric Research, the National Oceanographic and Atmospheric Administration (NOAA), the UK Meteorological Office and the University of California Los Angeles drove the research frontiers on climate change throughout the 1980s, and remain at the forefront of climate change research.

The first major international conference devoted exclusively to the problem of climate change was the World Climate Conference, organized by the World Meteorological Organization (WMO), and held in Geneva 1979. The conference gave rise to the World Climate Programme (WCP), a body jointly organized by the WMO, the United Nations Environment Programme (UNEP), and the International Council for Science (ICSU) (Agrawala, 1998). Throughout the first half of the 1980s there was a series of scientific workshops on climate change held in Villach (Austria) (in 1980, 1983 and 1985). At the last of these workshops a group of scientists issued a statement saying that 'in the first half of the next century a rise of global mean temperature could occur which is greater than any in man's (sic) history' (WMO, 1986, p1). They also suggested that there be a formal institution for international scientific collaboration to enhance understanding of the causes and impacts of, and solutions to, climate change. This latter recommendation was formalized in resolution nine of the tenth WMO Congress in May 1987, which requested that the WMO arrange a mechanism for such international scientific collaboration. With the backing of the US government and the support of UNEP the IPCC was established in 1988 (IPCC, 2004).

As scientific interest in climate change was developing and accelerating throughout the 1970s and 1980s, media interest was also increasing, and the foundations for a global climate change regime were being developed through a series of high-level meetings (beginning with the 1972 Stockholm Conference). In 1988 the first major international meeting of national policy makers on climate change was held in Toronto. This meeting was called 'The Changing Atmosphere: Implications for Global Security', and concluded that: 'humanity is conducting an unintended, uncontrolled, globally pervasive experiment whose ultimate consequences could be second only to a global nuclear war' (WMO, UNEP, and Environment Canada, 1989). There was a flurry of meetings among governments between 1988 and 1991, and by early 1991 formal negotiations for an international treaty on climate change were underway. These negotiations culminated in the United Nations Framework Convention on Climate Change (UNFCCC), which was signed by almost all countries at the 1992 United Nations Conference on Environment and Development in Rio de Janeiro (Paterson, 1996).

One of the remarkable features of climate change science is the extent to which it informs and is intertwined with policy. Since the 1970s climate change science and policy have co-evolved, with both scientific findings and policy decisions influencing each other (Hecht and Tirpak, 1995). That climate science has transferred its findings into a global policy regime is at the same time a remarkable achievement and strength, as well as a source of 'brittleness' as it leaves climate science open to a popular (if flawed) argument that its findings are less the product of research that conforms to the principles of science, and more the result of political imperatives (Demeritt, 2001). These critiques come from both the climate sceptics and those who are worried that the engagement has been between the (mostly) Organisation for Economic Co-operation and Development (OECD)-based scientists that construct GCMs and global scale policy processes such as the UNFCCC, and therefore climate change risks being a justification for management of the global commons by the North (Chaterjee and Finger, 1994). This latter critique carries more weight in the South, and is unsurprisingly nascent in the Pacific Islands given the way most scientific assessments treat the Pacific Islands as a homogenous environmental entity that is universally extremely vulnerable, and the way global institutions such as the UNFCCC and the Global Environment Facility alienate and marginalize many of the Pacific Islanders who interact with them.

International research programmes

The IPCC is the most prominent of the international scientific institutions that are associated with research on climate change, but there are many others, some of which predate the IPCC, and some of which interact with the IPCC in formal and informal ways. Many scientists are affiliated with more than one of these bodies, and this network of institutions and individuals means there is a socioepistemic network that is larger than any given institution, and which sustains an orthodoxy regarding problem identification, methods for assessment and approaches to solutions (Rogers and Marres, 2000; Janssen et al, 2006). In this section we examine those institutions affiliated under the Earth System Science Partnership (ESSP), as these are the most prominent and influential of the international scientific collaborative groups working on climate change and related environmental change issues.

The ESSP is a partnership formed in 2001 between four existing international research programmes: DIVERSITAS, the International Geosphere-Biosphere Programme (IGBP), the International Human Dimensions Programme on Global Environmental Change (IHDP), and the World Climate Research Programme (WCRP). The ESSP aims 'to undertake an integrated study of the Earth System', including 'its structure and functioning; the changes occurring to the System; [and] the implications of those changes for global and regional sustainability' (ESSP, 2009). Earth Systems Science is defined as 'the study of the

Earth System, with an emphasis on observing, understanding and predicting global environmental changes involving interactions between land, atmosphere, water, ice, biosphere, societies, technologies and economies' (ESSP, 2009). The predominant focus of ESSP is therefore on large-scale interactions among environmental systems. There is some consideration of the human dimensions of these changes in the work of the IHDP, which examines the social drivers of these changes (although this is largely about economic drivers), the implications of these changes for social (but again largely economic) systems, and institutional solutions to these changes.

There is an unclear division of labour within the four ESSP programmes, and the mandates of the IGBP and the ESSP are somewhat undifferentiated (ICSU-IGFA, 2008). The IGBP's research goals are to analyse: the interactive physical, chemical and biological processes that define Earth System dynamics; the changes that are occurring in these dynamics; and the role of human activities in relation to these changes. The goals potentially incorporate the specific concerns of DIVERSITAS (biodiversity), the IHDP (human dimensions) and the WCRP (climate change). However, DIVERSITAS also seeks to include human dimensions, and the IHDP is concerned with the social causes, impacts of and solutions to, all forms of environmental change, including climate change and biodiversity loss. These areas of overlap are a strength and not a weakness of the ESSP programmes, and are indeed essential given the multidimensional nature of sustainability issues. Drawing on expertise across the four programmes, the ESSP has launched four joint projects: the Global Carbon Project, the Global Environmental Change and Food Systems (GECAFS) Project, the Global Water System Project (GWSP), and the Global Environmental Change and Human Health (GEC&HH) Project.

DIVERSITAS, established in 1991, is centrally concerned with the changes in global biodiversity, and accordingly is closely linked with the United Nations Convention on Biological Diversity (CBD), which is an ex-officio member of the DIVERSITAS scientific committee. The DIVERSITAS Secretariat is hosted by the ICSU and is based in Paris. It is governed by a scientific committee of 11 members, of whom four come from outside the OECD, and three are female. There are 26 countries that have or are establishing DIVERSITAS committees, 12 of which are from non-OECD countries, but none of which is from a small island state. The human dimensions of biodiversity figure prominently in the bioSUSTAINABILITY project (one of four DIVERSITAS core projects). The steering committee of this project has eight members, four of whom come from outside the OECD, one of whom is female. DIVERSITAS receives most of its funds from various national funding agencies. In 2007, 15 national committees (12 from OECD countries) contributed two-thirds of its approximately US$700,000 annual budget (DIVERSITAS, 2007).

The IGBP was established in 1987 by the ICSU, and aims to provide scientific knowledge to improve sustainability. The IGBP Secretariat is hosted by the

Royal Swedish Academy of Sciences, and is based in Stockholm. It is governed by a scientific committee of 31 members, eight of whom come from outside the OECD, and ten of whom are female. The IGBP has an extensive network of national committees, including 47 committees from developing countries, four of which are from small island states, but none from the Pacific region. The IGBP has completed six, and has nine ongoing, core projects, many of which use models to advance understanding of interactions among processes in the earth system. Of these, the Global Land Project (GLP) and the Land-Ocean Interactions in the Coastal Zone (LOICZ) are jointly sponsored by the IHDP. On its website the IGBP reports that it is primarily funded by contributions from around 40 countries, and these contributions typically account for between 60 and 70 per cent of the average annual income of US$1.5 million.

The International Human Dimensions Programme on Global Environmental Change (IHDP) was established by the ICSU and the International Social Science Council (ISSC) in 1996. It aims to provide leadership on social science research to understand and respond to the challenges global environmental change presents to social systems, and to advance the interface between science and policy. The IHDP is guided by a scientific committee composed of 28 members, eight of whom come from countries outside of the OECD, and seven of whom are women. Its secretariat is based at the Bonn campus of the United Nations University (UNU). The IHDP has 32 National Committees and a further 30 national contact points. Of this total of 62 countries, 42 are from non OECD countries, and two are SIDS, including Fiji, which has a national contact point. The IHDP is sponsored by the ICSU, the ISSC and the UNU. In 2007 the funding for the IHDP Secretariat was US$1,134,000 although this does not include host-country funding for the secretariats for each of its six core science projects. Funding for the central IHDP secretariat largely comes from Germany (about 40 per cent) and the United States (about 20 per cent), with France and Spain being the next largest donors.

The World Climate Research Programme (WCRP) is the oldest and most well funded of the ESSP partners. It was established in 1980, and its main objectives are to determine the predictability of climate and the effect of human activities on climate. It now explicitly seeks to address the information needs of the UNFCCC, although the UNFCCC is not a member of the scientific committee as the United Nations Convention on Biological Diversity (UNCBD) is with DIVERSITAS. The WCRP is guided by a joint scientific committee, which has 18 members, six of whom come from countries outside the OECD, four of whom are female. Its administrative hub is located within the WMO in Geneva. The total budget of the WCRP and its activities are unclear, although its annual report for 2007–2008 shows that the combined income from the WMO, the ICSU, Intergovernmental Oceanographic Commission (IOC), plus some minor other sources totalled US$2,300,000 in 2007 (WCRP, 2008).

Across the four global environmental change research programmes that comprise the ESSP, representation from developing countries in governing bodies is modest, with between 25 and 36 per cent of committee members living in countries outside of the OECD. On none of these committees is there a representative from a Small Island Developing State. This is despite many programmes having national committees in developing countries (although it should be noted that we have categorized people by their listed country of residence, which means committee members who may have been born in developing countries, but who live in an OECD country, are counted as coming from the OECD). It is notable that there are no governing bodies that have a representative from a small island developing state, very few programmes have national committees in small island states, and from the Pacific only Fiji has a presence, and this is a national focal point (not a committee per se) for the IHDP.

There are three possible reasons for this under-representation of people from developing countries on national committees. First, it may be partly explained by funding: if it is not a requirement from donors that they have a member on a national committee, it is nevertheless strategic for programmes to include committee members from countries that provide funding. Thus, because the OECD countries dominate the funding of ESSP programmes, they also tend to be over-represented in their governing bodies. Kandlikar and Sagar (1999) and Runci (2007) argue that most of the funding for the IGBP and WCRP comes from national research programmes and is spent on researchers from the countries that provide the funding.

The second reason why developing countries are under-represented may be that, in as much as most of these programmes identify with the standard notion of natural science, and seek to maintain high scientific standards, they may tend to recruit committee members with well established research profiles, as measured by publications, citations, research funding and roles in other international bodies. The number of such scientists from developing countries is smaller than in OECD countries because their research budgets are smaller. Also when decisions are made about investing in science in developing countries, funding is usually directed towards immediate national priorities which often entail very applied research – such as research on improved cropping systems and the means to extend that knowledge to farmers – which means investment in research on longer-term and global scale issues is minimal (Kandlikar and Sagar, 1999; Runci, 2007).

A third reason is that the assumption of the scientific supremacy of the developed countries perpetuates itself in paternalistic patterns of collaboration. Glantz (2008, p22), for example, argues that a barrier to better participation in climate change research programmes is 'the North in that it continues to treat developing country scientists in general as junior partners always in need of more training'. Glantz, like Kanlidakar and Sagar (1999) and Runci (2007), notes that funding for South–South scientific collaboration is scarce, and that efforts at scientific collaboration among Southern countries are frequently compromised because 'usually the North is somehow involved and controlling the purse strings and, therefore, the agendas'. However, it is worth noting here that the IGBP is involved in the START (global change SysTem for Analysis, Research and Training) initiative, which is discussed below. None of these three factors particularly explain why women comprise between 22 per cent and 32 per cent of membership on national committees, which is instead most likely due to the underrepresentation of women in most fields of research in most countries.

The ESSP is perhaps best characterized as a network of networks. Its combined budget is very small indeed when compared to the national science budgets of most OECD countries. To compare, the United States Global Change Research Program (USGCRP) spent more than US$25 billion on climate research between 1989 and 2003 (Sarewitz and Pielke, 2007). This research, like that of the ESSP, has made profound gains in advancing awareness of the problems associated with environmental change, but has done little to help address changes in social structures to reduce the drivers of climatic changes, or to reduce the vulnerability of people to them. Recognition of this as a critical need in research is growing. For example, a recent review of the US National Research Council (NRC) Climate Change Science Program (CCSP) has concluded that 'the underlying human dimensions research needed to understand and develop sound adaptation strategies is a major gap in the CCSP ... a critical step in the process is for agencies with appropriate expertise to increase funding and take a leadership role in supporting, managing, and directing this research' (NRC, 2009, p7). However, as Pielke and Sarewitz (2003) warn, demands for new science such as this are generally not met with changes in supply, as existing institutions that supply science are slow or reluctant to respond for fear of losing funding to others who may have more claim to expertise in the new research priorities. In their words 'self interest trumps responsibility' (Pielke and Sarewitz, 2003, p27) with respect to the supply of research, and they argue this may actually have impeded effective decision making on solutions to climate change. The same concerns might be directed towards the ESSP, whose funding allocations suggest some reticence to shift research priorities in response to the needs of decision makers, particularly the needs of the 80 per cent of the earth's population who live outside of the OECD.

The Intergovernmental Panel on Climate Change

In 2007 the Noble Peace Prize was awarded jointly to the Intergovernmental Panel on Climate Change and to Al Gore Jr. for 'their efforts to build up and disseminate greater knowledge about man-made (sic) climate change, and to lay the foundations for the measures that are needed to counteract such change' (Nobel Foundation, 2007). The award was deserved, for without the work of the climate researchers involved with the IPCC – and this included both their research and their advocacy – society would be blind to the danger that it faces from climate change (Liberatore, 2001).

The IPCC was established in 1988 to provide independent scientific advice on the issue of climate change. It defines its role as being 'to assess on a comprehensive, objective, open and transparent basis the scientific, technical and socio-economic information relevant to understanding the scientific basis of risk of human-induced climate change, its potential impacts and options for adaptation and mitigation' (IPCC, 2004, pii). A unique feature of the IPCC is that it is an institution that seeks policy relevance, yet avoids being policy prescriptive. In this sense the researchers associated with the IPCC have a 'special assignment' as the IPCC moves their work beyond science and into the realm of science policy (Malnes, 2006). The IPCC produces assessment reports (of which there have now been four, the latest released in 2007), supplementary reports, special reports and technical papers. It also maintains a greenhouse gas inventories programme. The last two assessment reports each include the reports of three working groups: one on the 'scientific basis' of climate change, one on impacts, vulnerability and adaptation and one on mitigation.

Governments play an important and contentious role in the process that governs the preparation of all reports other than technical papers (the process for the assessment reports is shown in Figure 3.1). For the assessment reports, governments nominate scientific experts, review second order drafts, review the summaries for policy makers, and approve the reports of each Working Group. With respect to the influential *Synthesis Report* that emerges after each of the major assessments reports, and integrates findings across the three working groups, governments also review first order and revised drafts. At meetings at the plenary level of the governmental representatives to the IPCC they review and adopt the longer version of the Synthesis Report on a section by section basis, and review and adopt the summary version on a line by line basis – a process that can take many hours, at times lasting until dawn the day after the meeting commenced.

Governments also appoint the Chair of the IPCC Bureau, and they do intervene in this process, such as in 2002 when the United States vetoed the reappointment of Robert Watson because they believed he was too far beyond the influence of the US government (Haas, 2004). It has been suggested that this heavy involvement of governments is the result of developed country governments' attempts to influence the structure of the IPCC in order to regain control over the science process, which had hitherto been moving at a pace that governments could not keep up with (Haas, 2004). Given that governments are so heavily involved, it is remarkable that so many of the world's top researchers participate in the IPCC assessments, and the reports have significant credibility in scientific circles (Agrawala, 1998). This perhaps reflects the autonomy that researchers nevertheless do have in the IPCC process, and that many researchers

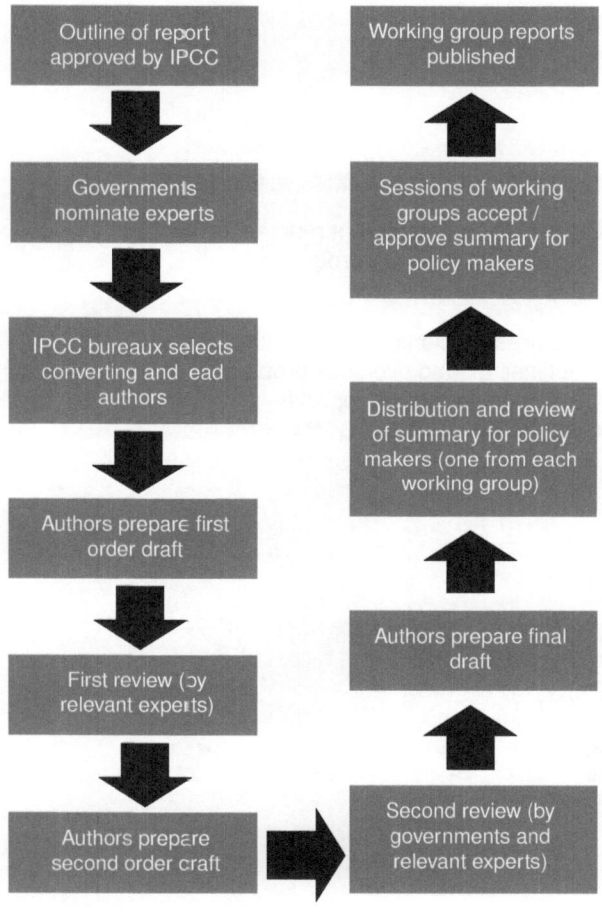

Figure 3.1 *Process for preparing and adopting the IPCC assessment reports*

have the sense that a slow and at times painful government review process, which can entail some circumscription of findings, is a trade-off worth making for the sake of the close engagement between governments and climate science as part of the IPCC process.

Both the IPCC and the UNFCCC are United Nations mandated bodies. In this respect the IPCC is both a scientific body and a formal governmental body (Moss, 1995). The First Assessment Report of the IPCC informed the construction of the UNFCCC. The Second Assessment Report (1995) was also used to inform the negotiating positions of some parties in the lead up to the Kyoto Protocol. Later assessment reports have been less well timed with respect to the UNFCCC processes.

It is important to recognize that unlike the ESSP programmes, the IPCC does not conduct any original research, with a notable exception being the development of the Special Report on Emissions Scenarios (SRES), which was organized

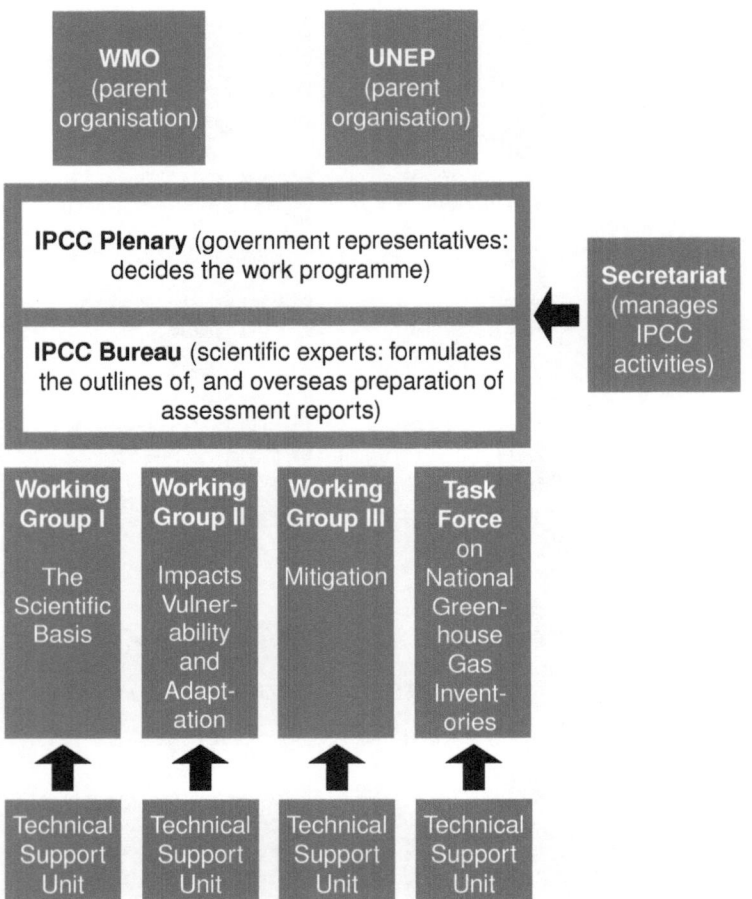

Figure 3.2 *Organization of the IPCC*

by the IPCC, and included the development of a work programme principally conducted by six scenario modelling groups whose work led to a set of storylines about possible future emissions pathways given potential trends in population, economic growth and technology (Nakicenovic et al, 2000). Apart from the SRES process, however, the IPCC's work is limited to assessing the latest literature on climate change, its impacts, and options for adaptation and mitigation. This does not mean that the IPCC has no influence in the nature of research on climate change. It arguably has a powerful influence through its identification of gaps in knowledge, and by valuing certain kinds of research over others in its assessment reports, thereby setting an agenda and creating incentives for certain kinds of research (Tonn, 2007; Dahan-Dalmedico, 2008).

More than any other scientific body, the IPCC has drawn attention to the risks that climate change poses to small island states. From its First Assessment Report in 1990 it has highlighted the risks that sea-level rises pose to low-lying coasts and settlements. These issues of coastal change as they relate to islands have been the domain of Working Group II (WGII) of the IPCC, which is the group responsible for assessments of the vulnerability of social and natural systems to climate change, and options for adapting to it (see Figure 3.2). Thus in the Second Assessment Report (1995) there was a chapter on Coastal Zones and Small Islands in the WGII report, in the Third Assessment Report (TAR) (2001) there was a chapter on Small Island States, and in the Fourth Assessment Report (2007) there was a chapter on Small Islands. In 1997 there was also a special report on the regional impacts of climate change, which included a chapter on Small Island States.

The IPCC's activities are funded by the IPCC Trust Fund, which is supplied by voluntary contributions. Funding is also received from the WMO, UNEP and the UNFCCC. The Trust Fund pays for the participation of researchers from developing countries in IPCC activities, whereas developed countries meet the costs of their researchers' involvement. The technical support units for each working group are funded by the governments that host them, and the IPCC Secretariat is hosted by the WMO. Trust Fund expenditure varies depending on the activities in a given year, but is in the order of US$5–7 million per annum (IPCC, 2005).

That the IPCC is an intergovernmental body does not imply that governments are equally influential in its processes and outcomes. Some developing countries, including many Pacific SIDS, do not send delegates to meetings of the panel. Within countries it is the IPCC focal point whose responsibility it is to receive correspondence from, and coordinate national activities in relation to, the IPCC (including, for example, sending delegates to plenary meetings, nominating experts and coordinating the review process). To be effective a focal point must: understand the unique characteristics of the IPCC as both an intergovernmental and scientific institution; have some sense of the breadth of the research that is covered in the IPCC reports; have some sense of the significance of various IPCC activities for the UNFCCC negotiations; and have a means to reach out to researchers in their country. This takes a uniquely skilled individual with good institutional support, and these people and resources are often not something governments in developing countries have at their disposal. Even India, a country with a large research base and a significant administrative capacity (relative to other developing countries) has at times had difficulty coordinating IPCC related activities (Kandlikar and Sagar, 1999). In the Pacific Islands individuals with the skills necessary to be an effective focal point are very scarce.

There is also very unequal representation of researchers from across the world in the IPCC's reports. Like the ESSP programmes, the majority of researchers participating in the IPCC reports are from developed countries. Table 3.1 shows that of the 393 convening lead authors, lead authors and review editors that participated in the Fourth Assessment Report (across all three Working Groups), two-thirds came from OECD countries (this does not include the 16 researchers from Russia), and one third came from outside of the OECD. Of this latter group, only 19 (3 per cent of all authors) came from Small Island Developing States, of which only two (0.3 per cent of all authors) came from the Pacific Islands. Both of the Pacific Island researchers were involved in the chapter on Small Islands in the WGII report, which had ten lead authors and editors, only one of whom was female.

Table 3.1 also shows that the WGI report has a higher proportion of authors from the OECD than the WGII and WGIII reports. When compared to the count of authors from developed countries in the TAR provided by Haas (2004), Table 3.1 shows that the proportion of lead authors from developing countries increased between the third and fourth assessment reports, which was also the case between the second and third assessment reports (Kandlikar and Sagar, 1999).

Table 3.1 *Developing country participation in the IPCC Fourth Assessment Report*

	OECD	Not OECD	All SIDS	Pacific SIDS
Working Group I	125	44	1	0
Working Group II	146	82	14	2
Working Group III	122	72	4	0
Total	393	198	19	2

The dominance of developed country researchers in the IPCC reports exists despite the IPCC Trust Fund helping to pay for the participation of developing country researchers in IPCC processes. The OECD dominance arguably under-mines the legitimacy of the IPCC in the eyes of some developing country decision makers (Kandlikar and Sagar, 1999; Haas, 2004; Lahsen, 2007; Runci, 2007), or at least enables them to cast doubts about the scientific basis for decisions that they find disagreeable. This questioning of legitimacy based in part on a lack of participation is less prevalent in the Pacific Islands, although there can be at times some resistance (or denial) of the more dramatic messages about climate risks that come from interpretations of its findings. Some in the region also have reserva-tions about the IPCC Special Report on Land Use, Land Use Change, and Forestry, and on Carbon Capture and Storage given that these are perceived to have legitimated carbon sequestration activities that have arguably weakened the environmental integrity of the UNFCCC and Kyoto Protocol.

There are many reasons for the under-representation of developing country researchers in the IPCC reports. These include the lower levels of investment in research in developing countries, and in particular in technology-intensive tools such as GCMs (hence the higher proportion of OECD researchers in the WGI report relative to the WGII and III reports). Of importance here is not just the volume of investment in research and development, but also the volume invested in public good research and development since much of the environmental science research that informs the IPCC is of a public good kind. Thus the more developing countries invest in public good research and development the more their researchers are represented in the IPCC reports. The huge increases in investment in research and development in China, which include increases in spending on the environmental sciences, is reflected in the 28 Chinese researchers listed as convening lead authors, lead authors or review editors in AR4.

Through the Subsidiary Body for Scientific and Technological Advice (SBSTA), the UNFCCC requests information from the IPCC that it considers necessary to guide negotiations. There is nevertheless no clear guidance in the UNFCCC on its relationship to the IPCC: the latter need not necessarily respond to requests from the former, the former need not consider the information from the latter. Nevertheless, in practice the two institutions are intertwined in many ways: many IPCC scientists sit on country delegations to the UNFCCC and some are active negotiators; and in negotiations on issues such as greenhouse gas stabilization targets, Land Use, Land Use Change and Forestry (LULUCF), and Carbon Capture and Storage (CCS), information provided by the IPCC frames the negotiations.

Thus, while it does not seek to recommend policies, the IPCC seeks to provide information relevant to decision making about climate change. This gives the IPCC tacit influence in policy, because information provision is central to the UNFCCC process and outcomes (Sabatier, 1988; Radaelli, 1995). It is this influence in climate policy that makes the perceived marginalization of developing countries from the IPCC decision making process, and from its assessments, troubling for the developing countries, making them feel 'captive' to the analysis of problems that emanates from the developed countries (Kandlikar and Sagar, 1999, p135). However, the majority of decision makers from the Pacific regard the IPCC favourably (although there has been some consternation regarding the LULUCF and CCS special reports).

The IPCC's goal of policy relevance also makes it easy for climate change sceptics to popularize their critiques by highlighting the difference between the IPCC and the idea of science as an objective and asocial process. Yet it has long been recognized that science is not free from social influences. Climate change is a problem characterized by high levels of uncertainty, urgency and risks of significant loss, and where decisions concern diverse and important interests, then the kind of post-normal scientific institution that the IPCC represents is necessary (Funtowicz and Ravetz, 1993; Saloranta, 2001). The critical issues are not whether the IPCC produces 'truth', given the ambiguity of this idea, but rather whether it produces quality science that is a reliable basis for decision making. Judged by this latter standard, the IPCC succeeds and is valuable, as recognized by the Nobel Prize Committee and all but a few researchers with a track record of publications in the field of climate change.

From the point of the view of the Pacific Islands, then, the IPCC is a positive institution: it has built an 'international epistemic community of scientists, policy-makers, and environmentalists united by their concern about climate change' (Demeritt, 2006, p472), which is important given the risks climate change poses to SIDS; and it has highlighted the unique and acute risks islands face from climate change. As a consequence of its work, there are few if any governments that are not aware that the decisions they make about the emissions of greenhouse gases have serious implications for the safety of small island states. Thus, the moral case for deep cuts in emissions that the SIDS make in the UNFCCC negotiations is one that is scientifically derived through the work of the IPCC.

The problems that some developing countries have with the IPCC concerning participation in governance and the assessment process are also felt by the Pacific SIDS – indeed, as we have shown, the Pacific SIDS are significantly underrepresented in the assessments, and few are active participants in the panel. However, because the findings of the IPCC reports are largely not threatening to the Pacific SIDS, these problems remain of minor concern most of the time. In terms of the way knowledge about climate change in the Pacific Islands is produced, because the IPCC does not produce research it is in some ways less a progenitor of knowledge and more and conveyor of it. Instead, what is known about climate change in the Pacific is a reflection of what the institutions that fund research seek to achieve. Given that there is very little endogenous funding of research in the region, this means that what is (and is not) known about climate change in the Pacific Islands is therefore the product of how developed countries and multilateral agencies support research on the region. The next chapter discusses the research initiatives relating to climate change that come from within the region (even if they are largely funded from money that comes from external sources).

Big science and big pictures

Its interest in the science of the earth system means the ESSP is concerned with large-scale processes, with the links between these and more localized processes and places being left largely to the IHDP. Earth systems science and its tools and techniques are strikingly panoptic. The big (earth systems) science is an all-seeing gaze from nowhere, assuming the mantle of objectivity, and it is communicated through powerful images and visualization technologies. It infrequently, if ever, considers particular groups of people, in places, and their actions and concerns, even though they too live in the earth system. Rarely is the research that is conducted inclusive of cases and issues from developing countries (Kandlikar and Sagar, 1999; Runci, 2007), and there are very few projects supported by any of the ESSP partners that include any of the Pacific Island countries in any substantial way.

Much of the research under the ESSP assumes 'an abstract global citizenry', which, by virtue of its production in developed countries, creates a vision of the world coloured by Northern norms, preferences and values (Kandlikar and Sagar, 1999, p130). Yet problems such as climate change are global justice problems and social problems as much, if not more than, they are global environmental management problems. This gets lost in earth systems science as its scale of resolution and its preferred methods preclude recognition of the people whose livelihoods and cultures depend directly on natural resources, which means they contribute almost nothing to the problem of climate change, but stand to lose the most due to their relatively higher exposure to risks and their lower capacity to adapt (see Adger et al, 2006). This justice perspective is central to the way people and leaders in the South understand and act on the problem of climate change, including in the UNFCCC. The approaches and findings of earth system science, therefore, do little to dispel the idea that international scientific collaboration, while addressing problems that affect many countries, does not include people from the majority of the world's countries, let alone give weight to their concerns. Thus while earth systems science rightly implores global response to global environmental problems, it has 'failed to engage society in creating the transformations that will lead to sustainability' because it does not consider the heterogeneity of interests and values across all societies, and the asymmetries of power that create barriers to change (O'Brien, 2006, p1).

Attending this global focus of the ESSP programmes is the dominance of projects informed by natural science approaches, and in particular models. Climate change is most often understood simply as being a problem for ecosystems and the environment itself, as if these values were themselves not socially produced, and there was some meaningful material distinction to be made between nature and culture. De-culturing and de-socializing environmental change research constrains policy as it makes problems seem like they occur in other places, and are caused by other people, which immediately says that it is others that are responsible for their solution (Sachs, 1993). Or, as is the case with analyses of the impacts of environmental change on GDP (Tol, 2002), it makes problems seem removed and less personally significant. Too rarely are the diverse array of individual and community needs, rights and values recognized in the kind of research the ESSP does, meaning that understanding of the causes of and solutions to environmental change is limited, and solutions are constrained by a lack of information.

Only the Global Environmental Change and Human Security Project (GECHS) of the IHDP attempts to link the larger scale processes to the needs, rights and values of individuals and communities, which drive environmental change, and in turn are affected by it. Despite this (or perhaps because of it), the GECHS project is one of the most poorly funded of the projects under the ESSP umbrella, suggesting that proponents of earth system science, and funders of it, have some way to go towards recognizing that the solutions to climate change will only arise through societal transformations, and will be poorly if at all served by further resolution of uncertainties about climate processes and impacts.

For its part, the IPCC and its assessment reports tend to reflect a global and an asocial framing of environmental change. The structure of the Working

Groups reflects the traditional approach to climate impacts research: it begins with modelling of atmospheric and oceanic circulation (WGI), focusing on the connections between the emissions of greenhouse gases, the resulting changes that GCMs consider to be likely, and their cascading affects on various biophysical systems and phenomena such as the coastal zone, water resources, agriculture and species distribution (in WGII). The effects of these changes on social systems are spread piecemeal throughout the WGII report, and in particular in the chapters outlining the regional impacts. The opportunities for, and constraints to, reducing emissions – which are very much social issues – are the subject of WGIII. The sequence of assessment is therefore along the assumed chain of causality in a linear manner: from the bench sciences through to the biological and earth sciences, ending with the social sciences (and at that largely with economics); and from corresponding global (atmospheric), through to regional (biophysical and economic regions of various kinds) and then more local scales (the places where people live) of assessment (Morgan and Dowlatabadi, 1996; Parson and Fisher-Vanden, 1997; Proctor, 1998). Because the assessment process is seen to begin with atmospheric systems, cascading changes through ecosystems, and only then consider impacts on social systems, the social is therefore almost always subordinate to the natural in climate change assessments, including those of the IPCC. In other words, social impacts and responses are seen to be *functions* of climate processes. In setting this structure the IPCC is implicitly valuing modelling approaches to research over other approaches, and is framing climate change as an environmental and to a lesser extent economic problem, rather than as a problem concerning people's needs, rights and values.

Mathematical and conceptual models dominate climate change research. The most powerful of these are the GCMs that simulate changes in the atmosphere and oceans resulting from increasing concentrations of CO_2. GCMs sit atop the climate change research hierarchy (Demeritt, 2001). Their power is at the same time historical, institutional and epistemological: they have been critical in creating awareness of climate change, they consume a major component of funding for climate change research, and they are the 'lynch-pin' (Shackley et al, 1998, p162) or 'hub' (Demeritt, 2001, p316) around which most other kinds of climate impacts research moves. GCMs are powerful because they are suitable and effective tools for projecting future climate, and because there is now an institutional commitment on the part of model makers and the majority of the climate impacts community to global-to-regional scale assessments and assessments of sectors (such as agriculture) based on inputting grid point data from GCMs into various models (for example of crop productivity) (Shackley et al, 1998; Demeritt, 2001). GCMs are also useful tools in answering the questions that have driven climate change research thus far: namely does climate change exist, how much of a problem is it and what are its consequences? (Pielke, 2005, p550). That impact models take data from GCMs and rely on them to set the parameters for their research establishes what Demeritt (2001, p320) calls a 'hierarchical distinction' between those who use the GCMs and the impact community, which implicitly includes even those who do research on impacts that are not so dependent on GCM outputs. GCMs also have some embedded cultural power as they are closely linked to computing power, and so most – and all the most influential ones – are produced in developed countries, the results of which are then delivered to developing countries (Henderson-Sellers and Braaf, 1996).

A widely used approach to assessing climate impacts is through the use of integrated assessment models (IAMs) that seek to integrate information through mathematical representations of aspects of natural and social systems (Risbey et al, 1996). The use of IAMs is testimony to the potential of GCMs to be a basis for the kinds of 'integrative science' (or at least integrative modelling) widely believed to be necessary for research into climate change. A number of observers have referred to the modelling approach to climate impacts as being to some degree 'deterministic' (Shackley et al, 1998; Agrawala, 2001; Demeritt, 2001) because they seek to determine the ways in which atmospheric processes transfer into changes in biophysical systems which in turn transfer into changes into social systems, and they implicitly treat climate change as the driver of social change to the exclusion of other environmental and social processes.

Thus far we have described the prevailing approach to climate impacts research that tends to see climate change as an environmental problem with somewhat separable human dimensions. Burton et al (2002) call this research 'first generation' climate impacts research and O'Brien et al (2004) describe it as focusing on 'outcomes' of climate change. There is, however, an alternative 'second generation' that is rapidly growing in influence and which is more firmly rooted in the traditions of social geography and its political-economy heritage. This second generation sees climate changes as 'a societal problem that has an environmental constituent' (Stehr and von Storch, 2005, p537), and as a 'broad development issue' (Burton et al, 2002, p151). This new approach seeks to analyse in detail the multiple sources of vulnerabilities and adaptive capacities at a local scale, but recognizing that these local manifestations are in part products of larger scale social forces.

This denaturalized (if not always interpretive) approach often takes the form of analysing vulnerability. The differences between first and second generation approaches can most clearly be seen when treating the issue of vulnerability: whereas the prevailing view sees vulnerability as being largely the product of exposure to changes in the physical environment, the alternative 'contextual' approach sees vulnerability as largely being the product of social forces (O'Brien et al, 2004). Whereas the prevailing view seeks to learn about future events, the alternative contextual view seeks to learn from historical and analogous events such as droughts, floods and cyclones, and what these events reveal about vulnerability and capacity to respond to climate change and variability (Glantz, 1988). One consequence of understanding vulnerability to climate change as a product of social context is that such research may reveal that climate change is not as important a problem to many communities as poverty, discrimination or various other direct and indirect violences and injustices. This does not mean climate change is irrelevant. It is likely to compound the adverse effects of these processes and is a risk to be considered; but it does mean that its solutions lie more in building safe and sustainable societies than has been recognized by climate researchers and policy makers.

Conclusions

This chapter has described some of the large research initiatives on climate change, and on climate impacts in particular, in order to explain something of the nature of climate research, the networks that produce it, and its links to policy. It has not argued that the existing body of research on climate change has been unhelpful or wasteful, indeed the accolades for the IPCC and its Nobel Prize are well deserved. Further, from the perspective of the Pacific Island countries, much good has come from climate change research. However, there are limits to this research endeavour, and some areas for improvement.

For as long as the supply of climate science is dominated by research institutions in the developed countries, and is dominated by researchers from those countries, it will be to various degrees mistrusted (for genuine or strategic reasons) as a basis for decision making by those in developing countries. Further, this supply is sometimes inefficient for the purposes of decision making in as much as it is not matched with demand, which is often more for assessments of what climate impacts mean for people at local levels than it is for further resolution of uncertainties in GCMs. We have also suggested that the extension of models cannot assist in meeting this demand for assessments of social vulnerability at local scales, which requires new approaches, and implies some decoupling of climate models from assessments of vulnerability and adaptation. More space for these new approaches to emerge is required. The most significant attempt to do this – the Assessments of Impacts and Adaptations to Climate Change (AIACC) project – is discussed in the following chapter, which explains the ways in which the institutions, standards and methods for climate change research discussed in this chapter have to some extent shaped the nature of Pacific-based climate research initiatives.

4

Pacific Science Initiatives

In the previous chapter, we discussed some of the main organizations involved in climate change research, and explained the kinds of approaches they use. We argued that these have not, and probably could not, produce much information of relevance to understanding the human dimensions of climate change in the Pacific Islands, and to informing decision making about adaptation. Given this, it is interesting to examine how knowledge about climate change in the Pacific Islands is produced, particularly given that there are some very bold pronouncements about the future of the region because of climate change (see Chapter 8 for examples).

In this chapter, we examine the production of science knowledge about climate change in the Pacific Islands that has ostensibly come from within the region. Our examination is guided by the questions: what science, by whom, for whom, and for what purpose? We argue that research on climate change in the region has been highly skewed towards the development of abstract models, and that this has significantly impeded the identification of social impacts and development of adaptive strategies.

The Association of South Pacific Environmental Institutions initiative

The Pacific Islands region has four universities and one scientific organization that deal with environmental issues. These are the Universities of Guam, Papua New Guinea and Samoa and the regional University of the South Pacific (USP) with its main campus based in Suva (and which has a campus in most other countries). There are of course smaller tertiary institutions that engage in research, such as the Fiji School of Medicine and the Papua New Guinea University of Technology, but these tend to have smaller, more sector based, and less environmentally-oriented research programmes. The South Pacific Applied Geoscience Commission (SOPAC) is the only other major institution that conducts research on environmental issues, although most countries in the region have meteorological services, and other regional organizations such as the Secretariat of the Pacific Community (SPC) and the Secretariat of the Pacific Regional Environment Programme (SPREP) have some involvement in research.

Researchers from the Pacific Islands were relatively quick to address issues of climate change and their implications for the region. A preliminary report was prepared by members of the Association of South Pacific Environmental Institutions (ASPEI), which first reported to SPREP in mid-1988. ASPEI, comprising the three major universities of the region – Guam, Papua New Guinea and USP (in Suva, Fiji) – responded promptly to the 1985 Villach meeting by putting forward a proposal, in 1986, to carry out an evaluation of the possible impacts of climate change in the Pacific region. Funding was made available from UNEP through SPREP, and in 1987 a working group was established to carry out the evaluation of impacts using scenarios developed by the UNEP Regional Seas Programme. The first preliminary report, which was presented at the second intergovernmental meeting of SPREP in mid-1988, included a regional overview and a number of case studies. Studies were based on sea-level rise scenarios used by the UNEP Regional Seas Programme of a 1 metre rise by 2050 and temperature scenarios building from a mean global warming of 0.5°C by 2000 through to 2.5°C by 2080 (ASPEI Task Team, 1988a). In all, ASPEI produced four compilations on climate change effects in the region, and material from three of these was included in a UNEP report published in 1990 (see Table 4.1).

The ASPEI work was a commendable effort given the little information available to the task team. The Pacific Islands were too small to have meaningful indicators of change from the general circulation models and, by and large, baseline data on the environment and human environment interrelations in the Pacific were, and to a large extent remain, limited. The first ASPEI report identified impacts on freshwater availability (especially in areas that experience long dry seasons), vegetation, soils and agriculture, health and comfort, coastal inundation, flooding and morphology, and on the incidence of tropical cyclones. These changes were in turn seen to have profound effects on the economies and societies of Pacific Island countries (PICs), including: the loss of agricultural production and the costs of bringing marginal lands into production to replace land that is lost to desertification or erosion; resettlement and out-migration from areas rendered uninhabitable; increased costs of engineered structures; and loss of cultural heritage. ASPEI stressed that its report was preliminary and considerably more work was required for a more definitive assessment to be possible. A second report was prepared for a meeting of a task team addressing the implications of climate change in the Mediterranean (ASPEI Task Team, 1988b). This was essentially the same as the first report but with two additional papers including a relative impact rating for PICs (see Table 4.1).

Table 4.1 *Reports by the Association of South Pacific Environmental Institutions, 1988–1990*

A. June 1988, Noumea	**UNEP REPORT[a]**
Preliminary Report	Overview (A)
PNG: Possible consequences of CC	PICs: Relative Impact Rating (B)
PNG: Agriculture in Highlands	PNG: Possible consequences, implications PICs (A)
PICs: SLR impacts	PICs: SLR, atolls and motu (C)
PNG: SLR selected sites	PICs: groundwater resources (C)
PNG: Deltaic floodplains	Tropical riverine lowlands (C)
PNG: Murik Lakes & Sepik Mouth	PICs: Plant ecophysiology (C)
SLR: raised coral and high islands	PICs: SLR on low coral islands (C)
PNG: Carteret atolls	PICs: Social and cultural impacts (C)
	PICs: SLR impacts (A)
B. October 1988, Split	Tongatapu (C)
PICs: Relative Impact Rating	PNG: SLR selected sites (A)
PICs: Overview of potential CC impacts	PNG: Deltaic floodplains (A)
PNG: Possible consequences, implications	SLR: raised coral and high islands (A)
PICs	PNG: Low-lying coasts (C)
PNG: Agriculture in Highlands	PNG: SLR Port Moresby and Lae (B)
PICs: SLR impacts	PNG: Murik Lakes & Sepik Mouth (A)
PNG: SLR Port Moresby and Lae	PNG: Carteret atolls (A)
PNG: SLR selected sites	PNG: Agriculture in Highlands (A)
PNG: Deltaic floodplains	
PNG: Murik Lakes & Sepik Mouth	
SLR: raised coral and high islands	
PNG: Carteret atolls	
C. July 1990, Majuro	**D. October 1990 Agriculture and Health**
PICs: SLR, atolls and motu	PICs: CO_2, CC and Crop physiology
PICs: groundwater resources	PICs: Impacts on agriculture
Tropical riverine lowlands	Atoll agriculture and ecology
PNG: Low-lying coasts	Coastal forestry
Tongatapu	Fiji: Socio-economic change in agriculture
PICs: Plant ecophysiology	Human thermal comfort
PICs: Social and cultural impacts	Health
PICs: SLR on low coral islands	

Notes: [a] (A), (B), (C) refer to original report that paper appeared in.

PNG = Papua New Guinea; PIC = Pacific Island country; SLR = sea-level rise; CC = climate change.

Many of the cautions in the initial report were not included in a layperson's version of it, entitled *A Climate of Crisis: Global Warming and the Island South Pacific* (Hulm, 1989). The first sub-heading in the Introduction is 'Innocent Victims' and the opening paragraph reads:

> *Global warming threatens the physical and cultural survival of several South Pacific societies. They are innocent victims of the northern hemisphere's 300-year orgy of fossil fuel burning in the name of industrialisation. (Hulm, 1989, p1)*

As we argued in Chapter 2, however, the Pacific Islands are not entirely innocent. Some people in the region have made poor decisions in relation to deforestation, especially in Papua New Guinea and Solomon Islands, consumption of fossil fuels in the region is increasing, and, like everywhere else in the world, corruption and mismanagement have led to environmental changes that increase vulnerability to climate change. As we discuss in Chapter 8, the discourse of innocence misrepresents the people of the region, and implies responses that do not take into account their circumstances and capabilities. Finding solutions to vulnerability to climate change in the region is unlikely to be aided by building a discourse of innocence that renders the region as passive and lacking agency.

A tone of alarm exists throughout the ASPEI booklet. For example, it suggested that 'a 4m rise in sea level would deal a death blow to several communities, produce a flood of refugees from other islands, devastate many societies through disruption of their crop production systems and social structure, and have a severe to catastrophic local impact in the rest' (Hulm, 1989, p2). This alarm arises in part because ASPEI used a scenario of 1 metre rise of sea level by 2050, which it said was conservative, raising the possibility of a 4.5 metre rise by 2100. As discussed in Chapter 1, none of the current estimates assume sea-level rise of this magnitude in the Pacific region, with the most credible upper estimate being a rise of 1.4 metres by the end of this century (Rahmstorf, 2007). Thus, while the reports by the ASPEI team were circumspect and replete with disclaimers regarding the certainty of their findings, the booklet painted the worst of pictures, no doubt because it was seeking to provide impetus for the industrialized countries to act to reduce emissions. Yet in so doing the booklet set a precedent for interpreting the findings of research in the most dramatic of ways. Most importantly, while the intended audience may have been decision makers in industrialized counties, in practice it was mostly read within the Pacific Island region, producing anxieties amongst Pacific Island leaders and their communities.

The ASPEI team reported to a special Intergovernmental Meeting on Climate Change and Sea Level Rise in the South Pacific in Majuro, in the Marshall Islands, in July 1989. The third report of the ASPEI task team was also presented to this meeting, together with the *Climate of Crisis* booklet and the first report (among other documents). The meeting concluded with a statement that said, among other things, that: 'the impacts of climate change are potentially catastrophic, threatening in the long term the very existence of low-lying countries'; the 'South Pacific people are innocent victims' (Pernetta and Hughes, 1989, p71). ASPEI produced a fourth report in 1990 reviewing the implications of global warming for agriculture, health and human comfort in Pacific Island countries (Hughes and McGregor, 1990). That same year a large volume was published by UNEP including material from the first, second and third ASPEI reports (Pernetta and Hughes, 1990).

The ASPEI reports did much to inform all of the governments in the region, together with Australia and New Zealand, about the possible effects of climate change in the PICs. While the scenarios were a little on the high side, they were in keeping with the thinking of the time. The work provided the region with essential information for policy makers and advisors, and played an important role in informing the approaches PIC governments took to international negotiations on climate change. However, it is unfortunate that the careful statements about risks in many of the background reports, which were based on detailed research, were lost in the *Climate of Crisis* booklet and the international reports that followed the booklet.

Since the ASPEI activities, research on climate change in the region has been of a far more piecemeal and ad hoc nature. There has not since been an initiative that is as comprehensive in geographical or substantive scope, and which includes as many of the regional research institutions. ASPEI stressed the need for better information about the likely effects of climate change and capacity of local communities to cope with it before the implementation of programmes. Despite this projects have nevertheless since been designed, and some implemented. The information deficit still remains nearly 20 years after it was identified.

Secretariat of the Pacific Regional Environment Programme initiatives

The focal point for climate change activities at the regional level has from the outset been SPREP – a regional intergovernmental organization with the mission of enabling and bringing about the sustainable management of Pacific islands' environments (note, prior to 2004 SPREP was known as the South Pacific Regional Environment Programme). It seeks to coordinate regional environmental initiatives, and provide assistance to its member states. Among its foci are issues such as ecosystem management, endangered species, waste management, environmental planning, international environmental agreements and climate change. SPREP is largely an administrative, political and policy oriented institution but it has been involved, often in an advisory and coordinating role, in a number of projects that have incorporated elements of research.

SPREP was a collaborating institution in some projects funded by the Government of Japan after the initial ASPEI project. These projects focused on integrated coastal zone management, coastal vulnerability and resilience in the context of climate change and sea-level rise. Studies were conducted at a variety of scales in Fiji, Samoa and Tuvalu, and incorporated physical science, engineering, geographic information systems and social science. They were primarily field-based assessments, and they provided important information on coastal exposure in the case study areas, in addition to developing methodologies for assessing coastal vulnerability and resilience, and classifying coastal areas accordingly (e.g. Nunn et al, 1994a, 1994b, 1996; Sem et al, 1996). This work led to the development of regional overviews (e.g. Mimura, 1999) and a partly qualitative vulnerability assessment methodology (e.g. Yamada et al, 1995). However, the methodology has not really been used since, in part because few such assessments have subsequently been attempted. At the same time the United States National Oceanic and Atmospheric Administration (NOAA) partnered with SPREP to conduct an assessment of vulnerability to sea-level rise in Majuro (Marshall Islands) (Holthus et al, 1992) and SPREP funded preparatory studies on the implications of sea-level rise, including in the Cook Islands (Sem and Underhill, 1992), Palau (Sem and Underhill, 1994), Samoa (Chase and Veitayaki, 1992), Tokelau (McLean and d'Aubert, 1993) and Tonga (Nunn and Waddell, 1992). These initial projects and reports made important steps towards better understanding the likely effects of climate change in parts of the Pacific region.

As part of the Pacific Islands Climate Change Assistance Programme (PICCAP), SPREP, in partnership with the International Global Change Institute (IGCI), and the United Nations Institute for Training and Research (UNITAR), developed a professional training course in climate change vulnerability and adaptation assessment for PICs. The course was developed and tested by IGCI in 1998, and was then modified and thereafter taught by USP. Central to the teaching of this course was the use of VANDACLIM, which is a simulation model to help train participants in integrated assessments (applied to the fictional island country of Vanda) (Warrick et al, 1999). VANDACLIM was developed by IGCI with funding from PICCAP. It and subsequent versions were based on the CLIMPACTS model that IGCI developed for New Zealand with the support of the New Zealand Foundation for Research Science and Technology (Kenny et al, 2000). UNITAR (Roncerel, 2002) reviewed the training course in their review of PICCAP and other training efforts, basing their assessment on PICCAP midterm review documents and a survey of 13 Pacific Island climate change team leaders. UNITAR (Roncerel, 2002) found that while the course had many benefits (discussed in Chapter 5), the modelling component was time consuming, added little value to the analysis of any given country, lacked detail on specific sectors and cases, was constrained by data availability, did not help assess adaptation options, and could easily become dated given changes in assessment methodologies. The report concluded that:

> *The modelling exercises based on an imaginary country, VANDA, are not necessarily any longer justified if not a considerable added value can be generated for the workshop participants' country. Therefore, this training package, including the software tool VANDACLIM, is most successfully applied when adapted and further developed to the regional or national circumstances, as was the case in the PICCAP project ... Still, in the V&A assessments established under PICCAP, rankings between sectors, or sector specific or case study vulnerability assessments, for instance, were lacking. And with regard to adaptation options additional needs have been assessed as being very large. (Roncerel, 2002, pp18–19)*

VANDACLIM was later tailored to make it more relevant to assessment of the impacts of climate change on the coastal zone, water resources, agriculture and health in Fiji and Kiribati (Hay, 2000), and then, with funding from the Government of New Zealand, a version called PACCLIM was developed (Hay, 2000). The Asia-Pacific Network for Global Change Research also provided US$68,000 for a workshop on PACCLIM conducted by IGCI in 1999. Along with other data, PACCLIM was used to inform the World Bank's (2000) study of the economic impacts of climate change in Kiribati and Fiji.

Yet PACCLIM is not mentioned in Kiribati's National Communication nor in its National Adaptation Plan of Action, both of which entailed assessments of vulnerability, perhaps because 'the lack or inadequacy of data on the environment of Kiribati limits the use of appropriate models on impacts and vulnerability' (Kiribati Government, 1999, p30). PACCLIM was also applied to the Cook Islands (COOKCLIM) to help assess vulnerability to climate change, although neither PACCLIM nor COOKCLIM seem to have influenced that country's national communication. As the Cook Islands' climate change officer has noted: 'although we have access to COOKCLIM and PACCLIM, the poor resolution of scale and focus on Rarotonga limits these models' applicability to the geographically diverse Cook Islands' (Carruthers, 2002, p5). A prototype of PACCLIM, called FIJICLIM, was also developed under PICCAP, and this informed the assessment of health risks in Fiji's National Communication to the UNFCCC.

The training in VANDACLIM provided under the auspices of PICCAP was intended to help the participants conduct assessments of vulnerability and adaptation in their countries, which was required as part of preparing the first of the national communications to the UNFCCC. It was intended that PACCLIM would be of assistance in conducting these assessments. Yet, of the ten national communications submitted by PICCAP countries, only Fiji explicitly made use of PACLIM or any other integrated assessment model (IAM). PACCLIM and its variants have been identified by countries as being of little use, because they lack socio-economic data, are very time consuming, create dependence on external 'experts' for support, do not help engage decision makers, and do not lead to information that can help with planning adaptation (as recounted to us, but see also Carruthers, 2008). Thus countries in the region have 'very little faith' in these models (Nakalevu, 2007). That the models have limited utility had been at least informally identified by most countries in the region some time before the Assessments of Impacts and Adaptations to Climate Change (AIACC) project, discussed below, was implemented in the region.

START-Oceania (including APN and AIACC)

Throughout the 1990s, the predominance of scientists from developed countries in international scientific bodies working on global environmental change – such as the ESSP partner programmes and the IPCC – became increasingly obvious. At the same time it was becoming more evident that the developing countries would bear the brunt of many of the effects of global environmental change. Accordingly, under the sponsorship of the International Geosphere-Biosphere Programme initially, and later with support from the International Human Dimensions Programme on Global Environmental Change and the World Climate Research Programme, the SysTem for Analysis, Research and Training (START) was established in 1992 to build capacity in, and knowledge about, environmental change in the developing world. START aims to build networks of collaborating researchers and institutes to conduct research, provide information to decision makers, and build scientific capacity in developing countries. START is sponsored by the ESSP, although its role as a capacity building institution within the ESSP is somewhat restricted (ICSU-IGFA, 2008).

The international START secretariat is located in Washington DC, and is funded by the US Climate Change Science Program and the US National Science Foundation. It has five regional centres: in Dar es Salaam (for the Africa regional network); Beijing (for the Temperate East Asia regional network); Bangkok (for the Southeast Asia regional network); New Delhi (for the South Asia regional network); and Suva (for the Oceania regional network). The coordinating office for the Mediterranean network is in Toulouse, and there are three regional nodes: in Accra, Cape Town and Nairobi. These centres, offices and nodes are funded by the governments that host them. In addition to this core funding, START receives grants from various agencies, including the Asia Pacific Network for Global Change Research (APN), the Canadian International Development Agency (CIDA), the Department for International Development (UK), European Science Foundation (ESF), the German Federal Ministry for Education and Research, the International Council for Science (ICSU), the MacArthur Foundation, the National Science Council of Taiwan, the Swedish International Development Cooperation Agency, the Research Council of Norway, the EC-EuropeAid Co-operation Office and the Global Environmental Facility (GEF).

As mentioned above, one of START's regional networks covers Oceania, and its secretariat is based at USP in Suva (Fiji). Compared to the other START regional networks the Oceania one is the least developed, which is not surprising given the small number of educational and scientific institutions in the region. Nevertheless START-Oceania has facilitated research projects in the region, largely with funding from APN.

Much of the APN funded research throughout the Pacific is climate change related, although other environmental change issues are also supported. There are two main categories of APN funding (grants are typically in the order of US$40,000). The first is through an annual regional call for proposals, and the second is for projects that contribute to capacity building (called CAPABLE grants). Between 1997/98 and 2008/09 218 projects were funded through the annual contestable fund, and of these 11 (5 per cent) were or are directly related to the Pacific Islands. Of approximately 60 CAPABLE projects thus far, seven (12 per cent) have been awarded to Pacific Island projects, mostly to USP. While this contribution to the Pacific is relatively small, the APN region also includes most of Asia, and close to half of the global population.

The funding of projects in the Pacific has been very useful in sustaining some momentum for environmental change research in the Pacific Islands, and in meeting emerging knowledge needs. Moreover, several of the APN projects have facilitated research on the human dimensions of climate change in the Pacific, and in particular on adaptation, and thus have helped meet a key research need. Nevertheless, the research in the Pacific that has been funded by APN tends to be ad hoc, reflecting the interests of those applying for grants, and there is a pressing need for a coordinated regional research strategy that can guide all research funding providers, and which will build a more coordinated understanding of climate change, its effects, and options for responses.

One of START's key and most successful activities has been the AIACC research programme. AIACC aimed to: enhance understanding of vulnerability and adaptation to climate change in developing countries; grow the capacity of developing countries to conduct climate change vulnerability and adaptation assessments; and to inform climate change adaptation planning and action (Leary and Kulkarni, 2007). The project was a collaborative effort: it was developed by the IPCC, funded by the GEF (which provided US$7.5 million), implemented by UNEP, and executed jointly by START and the Third World Academy of Sciences (TWAS). An additional US$475,000 came from the United States Agency for International Development (USAID), CIDA, the Rockefeller Foundation, the US Environment Protection Agency, the World Bank, and the in-kind funding from participating institutions was valued at US$1.8 million.

The project ran from 2001 to 2007, and there were 24 AIACC projects: 11 in Africa, five in Asia, five in Latin America and three in small islands (one of which was in the Pacific). Projects were selected via a competitive process based on technical merit. They were intended to mark an important change in direction from the original (or 'first generation') top-down climate impact assessments to ones in which stakeholder needs were incorporated and a better understanding of vulnerability and adaptation was the goal (AIACC, 2002). The project involved more than 350 researchers from 150 institutions in 50 developing countries, and 12 developed countries were involved. Thirty of these researchers were authors of the IPCC Fourth Assessment Report. It produced two books (Leary et al, 2007a; Leary et al, 2007b), more than 60 peer-reviewed papers and chapters, over 40 peer-reviewed on-line working papers, and 25 student theses. AIACC publications were cited 102 times in the report from Working Group II of the IPCC Fourth Assessment Report.

The Pacific AIACC project was administered by USP, and the participating stakeholder institutions were IGCI in New Zealand, the Fiji Department of Environment and the Cook Islands Environment Service. SPREP played an important role in screening and supporting various proposals for the Pacific AIACC project, and was subsequently involved in governance of the project. The project received US$220,000 from AIACC (Koshy, 2007). It was focused on advancing existing AIMs. The main aim of the project was to incorporate human dimensions into existing IAMs, which had mainly focused on identifying biophysical effects such as coastal erosion and changes in crop production. However, unlike the aims of other second generation approaches to adaptation where stakeholder involvement is intended to take a prominent role, the Pacific AIACC project, by the very nature of its modelling approach, reinforced a top-down methodology (and was therefore somewhat contrary to the aims of AIACC).

The project had three specific aims: to develop the next generation of integrated assessment methods and models; to expand understanding of climate impacts and adaptation by implementing testing and applying the methods and model in study areas; and to build in-country capacity to use these advanced methods and models (Koshy, 2007). The study areas for testing the models were Natadola and Navua in Viti Levu (a high island in Fiji), and Aitutaki in the Cook Islands. Aitutaki is described in the final report as being 'typical of the myriad [of] low-lying atoll islands of the Pacific' (Koshy, 2007, pxii) but should more properly be described as an 'almost atoll' with only a small remnant of the original volcanic island and a wide lagoon and barrier reef islands. Given that the vast majority of the inhabitants of Aitutaki live on the high island, it is not representative of most atolls.

The emphasis of the project was very much on the development of the 'next generation' of IAM for small islands. This model – SimCLIM – was the next iteration of the country-applied models of PACCLIM (such as FIJICLIM) developed during PICCAP. Informing adaptation was not a priority (Koshy, 2007, p35), which is reflected in the summary of findings in Leary and Kulkarni (2007, pp142–143), that includes 44 lines about the successful development of the next generation IAM, and only eight lines on lessons learned about vulnerability and adaptation in the region. Progress reporting (Nunn, 2003a, 2003b) makes it clear that the principal function of USP in the project was to 'collect data which can be incorporated into the model being developed by IGCI', with the 'emphasis ... on data collection, data screening, and onward transmission to the modellers at IGCI' (Nunn 2003a, p2). IGCI, for their part, reported 'continual delays in receiving data required for modelling purposes' (Nunn, 2003a, p4).

The final report for the project notes that a 'serious setback' to the project was a 'paucity of relevant data and information' Koshy (2007, ppviii, 17). This presupposes that the only reason to collect data is to inform the development of models, and that the standard by which data should be judged is its ability to do this. It is unfortunate that a project about the region ends up highlighting data inadequacies in this way, particularly when that data was to be collected by a regional research institution, and the standards for that judgement is only whether or not the data is suitable for the development of a model. This devalues the large amounts of data collected about social and environmental conditions in the region through the efforts of regional and national agencies such as USP, SOPAC, SPC, national statistics offices, national health departments and national environmental agencies. It also ignores the large amounts of data that could be collected to support this existing body of knowledge, which may come from local people and could meaningfully be used to assess vulnerability and adaptation to climate change in the region.

The Pacific AIACC project had modest outcomes in terms of capacity building. It employed two Fijian research assistants, and data collection was conducted by 12 undergraduate students from within the region. Four postgraduate students from USP were also employed in the project, although none is reported to have completed a research thesis. The project introduced 40 individuals from PICs to SimCLIM (Koshy, 2007). While a cadre of Pacific civil servants may have been introduced to SimCLIM, the communities who will have to adapt were not included. The project contributed to one research thesis – by a Masters student enrolled at the University of Waikato in New Zealand. The project also assisted in the development of a training version of SimCLIM, called TrainClim, that has been incorporated into the V&A training offered by USP.

The Pacific project produced one journal article reporting on climate trends in Viti Levu (Mataki et al, 2006a), one AIACC working paper reporting on a field-based and largely socio-economic study of adaptation in one of the project's three study areas of Navua (Mataki et al, 2006b), and one book chapter discussing ways to harmonize local and national efforts to promote adaptation (Mataki et al, 2007). Despite the project's aim that the study areas were to be used to collect data to inform the model, the published studies are not reported in this way, and they make almost no reference to SimCLIM. SimCLIM is itself an output of the project, which is now marketed by CLIMsystems (CLIMsystems, n.d.).

We do not wish to give the impression that SimCLIM is an unhelpful product, and there has been some positive feedback from people introduced to it. It has a role amongst other approaches to climate change impact assessment and adaptation planning. However, it continues an emphasis on the use of commercial models developed by groups outside of the region for assessing vulnerability, and which can only provide limited help to inform adaptation. The emphasis of the Pacific AIACC project on the development of a model was contrary to the recognition by most climate change decision makers in the region that models were of limited utility for decision making; indeed it was somewhat contrary to the aims of the AIACC project itself. It was also a goal contradicted by the published outputs of the study, most of which were about adaptation informed by bottom-up field-based assessments, and to which SimCLIM contributed relatively little new knowledge.

Lacunae

It is instructive to compare statements about the state of knowledge about climate change in small island states between the Third (2001) and Fourth (2007) Assessment Reports of the IPCC. Chapter 17 of the TAR observed that:

> there is some uneasiness in the small island states about perceived over reliance on the use of outputs from climate models as a basis for planning risk reduction and adaptation to climate change. There is a perception that insufficient resources are being allocated to relevant empirical research and observation in small islands. Climate models are simplifications of very complex natural systems; they are severely limited in their ability to project changes at small spatial scales, although they are becoming increasingly reliable for identifying general trends. In the face of these concerns, therefore, it would seem that the needs of small island states can best be accommodated by a balanced approach that combines the outputs of downscaled models with analyses from empirical research and observation undertaken in these countries. (Nurse and Sem, 2001, p870)

Six years later, Chapter 16 of the AR4 noted that:

> momentum for vulnerability and impact studies appears to have declined ... Our assessment has identified several key areas and gaps that are under-represented in contemporary research on the impacts of climate change on small islands ... In contrast to the other regions in this assessment, there is an absence of demographic and socio-economic scenarios and projections for small islands ... assessing the impacts of climate change on the human systems of small islands remains a challenge ... the understanding of adaptive capacity and adaptation options is still at an early stage of development ... [there is a] need for further research, including the assessment of practical outcomes that enhance adaptive capacity and resilience. (Mimura et al, 2007, pp711–712)

So, despite longstanding recognition that the Pacific Islands are very vulnerable to climate change, there have been few locally oriented empirical studies of vulnerability and adaptation in the region. A lack of cognate data or methodological guidance does not explain these lacunae: there are good observations of change in climate and ecosystems (e.g. Salinger, 2001); there is a wealth of local knowledge to harness; there are methodologies available (e.g. Lim and Spanger-Siegfried, 2004); and there are many analogous events from which to learn (Glantz, 1988). Local-level studies are not more expensive than modelling studies either. For example much could have been done to advance understanding of vulnerability and adaptation in the region, including training many higher degree research students, with the funding provided to the AIACC project.

IAMs were used in PICCAP and AIACC in part because they integrate conceptually and mathematically with the science of Atmosphere-Ocean GCMs, and in part because their advocates were often entrepreneurial in negotiating research funding. It has also been the case that scientific organizations with modelling capacity and the capacity to win funding have developed working relationships with regional partners for the sake of complying with funding rules that specify the inclusion of regional partners, rewarding the local partners with some modest inclusion in projects and their funding, and inducing their support for future bids for funding. In this way individuals and institutions in the region continued to advocate for modelling based studies that only organizations outside of the region could conduct. The possibility of funding for alternative projects and approaches was stifled by the hegemony of the modelling agenda and its institutions and their regional clients. In articulating the distinction between these earlier studies and the subsequent CBDAMPIC project (see Chapter 6), Nakalevu (2006, pp10–11) writes that four PICs involved in the project did not support further integrated modelling because the models are unable 'to do anything else apart from what had been programmed', and their use would continue a situation where countries are 'dependent on international centres that have these technologies and also pay huge sums of money to access them' so that 'the end result would have been very little progress taking place at the community level'.

Thus 'experts' and their top-down generic models have tended to set the norms for knowledge about climate change in the region. In doing so they have marginalized the value of, and crowded-out opportunities for, local knowledge and approaches. Whenever data and/or technology have not been available, the language of modelling science has not been comprehensible, or when the utility of generic models has been questioned, the response has been that the deficiencies are those of the islands (Barnett, 2010). Yet it may be that people in the region cannot engage with models and the assumptions of the world that underpin them. As we saw in Chapter 2, in the Pacific, nature is indivisible from the social, and the economic and cultural and political are indivisible from genealogy (Kempf, 1999), whereas models homogenize people, see social life as the sum of rational individual actions, assume 'culture' is separable from other aspects of society, and assume that nature and society are independent 'facts' (Proctor, 1998; Shackley and Gough, 2002). Models therefore tend to produce knowledge of the Pacific Islands that is alien to, and alienates, people in the region (Barnett, 2010). This is not to say that models do not have a role, but it is to say that their hegemony in climate-impacts research limits understanding. They should be one important element of a research endeavour characterized by methodological pluralism, and innovation driven by constructive engagement among proponents of different methods presenting evidence from studies of diverse places and cases.

Some decolonization of climate-impacts research is clearly required. There are new approaches for achieving this, called 'second generation', 'vulnerability/ adaptation' or 'bottom-up' approaches. These include the United Nations Development Program's Adaptation Policy Framework (Lim and Spanger-Siegfried, 2004), the National Adaptation Plan of Action Guidelines (UNFCCC, 2002), and the method used in the CBDAMPIC project (discussed in Chapter 6). These approaches offer a framework of linked concepts, questions, methods and principles for assessment that can be combined in various ways to suit particular research tasks. They diverge from earlier methodologies because they focus more on current vulnerability to present climate risks in order to learn about future risks; focus on smaller scales and at scales where decisions about adaptation are made; prioritize social systems and highlight the social and economic forces that create vulnerability; seek to inform decisions about adaptation; emphasize the problem of climate extremes (not just changes in mean conditions); include stakeholders in assessments; integrate a wider range of existing data (for example from studies of resource management, planning, economic development, household expenditure, and decision-making processes); and consider the capacity of social systems to implement adaptation actions.

These approaches are better suited to PICs. They seek information that people already possess, value traditional knowledge and transfer the capacity to conduct assessments to countries. They help give countries confidence in, and ownership of, the results of assessments and their proposed adaptation actions; contextualize and communicate vulnerability in the particular contexts in which it arises; determine effective and legitimate adaptation strategies that are locally suitable; and empower Pacific Islanders to 'own' what is known about climate change in their places. In Chapter 6 we discuss some projects that have achieved many of these outcomes.

Conclusions

Much of the knowledge about climate change in the Pacific Islands is not produced by Pacific Islanders. This is in part for financial reasons, as regional organizations and countries do not in themselves have much money to fund research (although this is more of constraint in the case of modelling type research which tends to have higher costs). Thus research institutions rely on external sources for funding, and on partnerships with external institutions that have the wherewithal to bid for funding. This dependence means that research on climate change in the region is often dictated more by supply than demand. It also means that the kind of research that is done relies on the skills of researchers from outside the region. More problematically, rarely does this research adequately value the knowledge that people living in the region have about climate, the effect of climate variability and change on their lives and livelihoods, and ways to adapt. Because knowledge about climate change in the Pacific is rarely produced by

Pacific Islanders, it rarely serves the purposes that decision makers in the region think it should, and it rarely takes account of the knowledge, needs, rights and values of the people who will be exposed to climate change, and who will have to adapt to it.

As this chapter has shown, this problem of dependence on external actors was less the case in early research efforts, where donors (and in particular the Japanese government) were funding field-based studies conducted by teams composed of researchers from within and outside of the region working in genuine partnership. These studies laid a foundation for research management and research capacity development, and for the transfer of knowledge from research to decision making that was somewhat squandered in later years. The problem of dependence on external actors emerged with the use of models as the dominant approach to research, as these rely on a few experts based outside of the region, and they cannot include much of the field-based data that best reflects local concerns. These models became entrenched under the PICCAP project.

There was from the outset some disquiet within the region about – and resistance to – the use of models and dependence on the expertise of modellers. Yet the ability of people in the region to articulate these concerns was limited by their inexperience with research relative to those providing the tools, their largely non-confrontational nature, a lack of choice between approaches, and the nascent patron–client relationships that formed between some organizations involved in this research enterprise. Further, even if it has been less well funded and not so highly promoted, there has always been good work producing, communicating and transferring knowledge about climate change done by Pacific Island scholars such as Pene Lefale, Melchior Mataki, Taito Nakalevu, Graham Sem and Joeli Veitayaki, USP-based researchers such as Bill Aalbersberg, Kanayathu Koshy, Patrick Nunn and James Terry, earlier contributors to the ASPEI reports, and the efforts of national climate change coordinators and their teams in their reporting to the UNFCCC.

With the passage of time countries and researchers in the South Pacific have noted the limitations of the modelling tools that have been applied in the region, and the inequities associated with research funding intended for the region being spent on models and institutions based outside the region. Nevertheless, the problem has long been that the funding of models has come at the expense of research that could have built local capacity, engaged local communities, sought to inform decisions about adaptation at the scales at which such decisions will be made, and highlighted that climate change is a social more than it is an environmental problem. In recent years the desire in the region for research of this kind has been well communicated to donors and external research agencies, and as Chapter 6 shows, in response some innovative projects on adaptation that include research as an element have emerged.

5

The Architecture of Climate Change Policy

In this chapter, we explore the ways in which the global climate change policy process influences climate change activities in the Pacific Islands, and the way the Pacific Islands themselves influence the global climate change policy process. The United Nations Framework Convention on Climate Change (UNFCCC) is critical for the Pacific Islands because it is still widely regarded as the institution most likely to lead to reductions of greenhouse gases. Although few people think the UNFCCC can achieve reductions in emissions to a level to avoid significant climate impacts in the Pacific region, in theory every reduction in emissions increases the time the Pacific Islands have to adapt to climate change. Insight into the UNFCCC is therefore important for understanding the issue of climate change in the Pacific. Further, as suggested in Chapter 3, decision making under the UNFCCC is to some degree influenced by research on climate change, and as we show in subsequent chapters the decisions and processes implemented by the UNFCCC have some influence on climate change policy and planning, including in the Pacific Islands. Thus the UNFCCC is a critical part of the science-policy process that shapes responses to climate change in the Pacific.

This chapter begins by outlining the emergence of the UNFCCC. It then discusses the UNFCCC in some detail, including its main financial mechanism, the Global Environment Facility (GEF). It then discusses the role of some of the main actors in the UNFCCC in regards to the Pacific Islands, including the Alliance of Small Island States (AOSIS), the Pacific Islands Forum, and the developed countries that engage most closely with the region on climate change and other issues.

The emergence of climate change as a global policy issue

As discussed in Chapter 3, in the 1970s uncertainties about the nature of climate change led to an expansion of research into the phenomenon (Weart, 2008). By the end of the decade the First World Climate Conference was held and the participants concluded that it was likely that the climate may be altered by increasing emissions of CO_2, although this had almost no effect on public perceptions of the problem. It wasn't until later in the 1980s that significant political concern began to emerge (Weart, 2008). The Villach workshops marked a shift in impetus for political intervention, arguing that 'the rate and degree of future warming could be profoundly affected by governmental policies on energy conservation, use of fossil fuels, and the emission of some greenhouse gases' (UNEP/WMO/ICSU Conference, 1986, pxx).

Concerns about the global atmosphere soon resulted in international governmental action. However, the first international agreement was focused on stratospheric ozone depletion and the role of chlorofluorocarbons (CFCs) and halons in this process, and not climate change. Concern about ozone was embodied in the Vienna Convention (1985), and targets for reductions of harmful gases were set in the subsequent Montreal Protocol (1987). In 1987 the World Commission on Environment and Development (WCED, 1987) published *Our Common Future* (The Brundtland Report) which established the notion of sustainable development. The report drew attention to the findings of the Villach meeting and to the issue of climate change.

Also in 1987, the President of the Maldives, Maumoon Gayoom, addressed the Commonwealth Heads of Government Meeting (CHOGM) in Vancouver, Canada. He voiced his concern over the dangers that climate change would impose on his atoll country. A number of Pacific Island leaders were present and in the following year brought their concerns to the annual Pacific Island Forum leaders meeting. Later in 1988, the New Zealand Deputy Prime Minister (and Minister for the Environment) gave a detailed lecture in Hawaii on the implications of climate change for the countries of the Pacific region (Palmer, 1988). Following the CHOGM, the Commonwealth Secretariat formed a Commonwealth Group of Experts to examine the implications of climate change for its member countries. The group reported in September 1989, noting that there was special need for concern for low-lying island countries (Holdgate et al, 1989).

Also in 1988, the Toronto Conference on The Changing Atmosphere: Implications for Global Security was held. Its conclusions were hardly equivocal, with the final statement asserting that global warming was 'an unintended, uncontrolled, globally pervasive experiment whose ultimate consequences could be second only to a global nuclear war', and it called on 'governments to work with urgency towards an Action Plan for the Protection of the Atmosphere' (WMO et al, 1989, p292).

The establishment of the Intergovernmental Panel on Climate Change (IPCC) in 1988 provided governments with a body to inform them of developments in climate science and understanding of global warming. The first IPCC assessment, initiated in 1988, included Working Group III, the 'Response Strategies Working Group' (RSWG), which included in turn a subgroup working on

Coastal Zone Management (CZM). In 1990, a meeting was held in Perth, Western Australia, by the CZM subgroup to ensure that tropical states with coastlines, including islands, were enabled to have input into the CZM report of Working Group III. Seven Pacific Island countries (PICs) were represented at the meeting: Federated States of Micronesia, Fiji, Kiribati, Papua New Guinea, Tonga, Tuvalu and Samoa.

That same year (1990), Tuvalu emerged in its role as climate change *cause célèbre* when the Prime Minister, Bikenibeu Paeniu told a session of the Second World Climate Conference that included, among other leaders, Prime Ministers Thatcher from the UK and Rocard from France, that:

> *I can assure each and every one of you that I speak today from real experience because I live on one of the … smallest island groups in the Pacific. We are therefore, along with others, extremely vulnerable to environmental hazards and the dangers of the Greenhouse Effect and sea level rise. These are the problems which we have done the least to create but now threaten the very heart of our existence.*

The international media then began to report the story of a small and vulnerable piece of paradise at risk because of the actions of the large and the powerful. Prime Minister Paeniu continued to lobby countries with large emissions profiles, becoming one of the most influential people from the region in bringing the implications of climate change for small island states to the attention of the world. Climate change, or at least its impacts, was soon to become identified with the small and apparently vulnerable nations of the Pacific.

Following the publication of the IPCC First Assessment Report, in December 1990, the UN General Assembly established an Intergovernmental Negotiating Committee on the Framework Convention for Climate Change (INC/FCCC). This set the platform for the development of an international agreement on climate change, which was scheduled to be finalized at the United Nations Conference on Environment and Development (UNCED), also known as the Rio Earth Summit. The UNFCCC was ready for signing in Rio in 1992.

The UNFCCC and the Kyoto Protocol

The UNFCCC and its 1997 Kyoto Protocol are complex institutions: they are the product of over 15 years of laborious and contentious negotiations among over 150 parties. The configuration of the UNFCCC mirrors that of the Vienna Convention for the Protection of the Ozone Layer in that it requires very little from its members apart from reporting requirements and general commitments to protect the global atmosphere (O'Riordan et al, 1998). But the UNFCCC has the potential for parties to the convention to develop stronger commitments. In the case of ozone this came with the Montreal Protocol, which has been adjusted several times to increase the commitments of parties to reduce emissions of ozone-depleting substances. Widely seen as a success story, the Montreal Protocol was assisted by technological developments that enabled alternatives for many of the ozone-depleting substances to be used and caused relatively little disruption to the industries, such as refrigeration, that used them (Anderson et al, 2002). In comparison, the UNFCCC and its first protocol that commits some countries to emissions reductions targets (the 1997 Kyoto Protocol) has been far less successful in reducing emissions of gases because there are not as many easy technological fixes to reduce emissions. Accordingly, achieving significant reductions will require considerable investment, and social and cultural change among the world's growing wealthy and middle classes.

The preambular text to the UNFCCC recognizes that small island countries are among those that are particularly vulnerable to the adverse effects of climate change. Yet, the convention's main objective is to achieve:

> *stabilization of greenhouse gas concentrations in the atmosphere at a level that would prevent dangerous anthropogenic interference with the climate system. Such a level should be achieved within a time frame sufficient to allow ecosystems to adapt naturally to climate change, to ensure that food production is not threatened and to enable economic development to proceed in a sustainable manner. (Article 2)*

Exactly what constitutes dangerous climate change is a matter of some debate, and is somewhat relative to the perception of climate risks that different groups have (Dessai et al, 2004). Further, were it possible for parties to agree on some more precise benchmarks of unacceptable outcomes, the level of concentration of greenhouse gases in the atmosphere that would cause those outcomes would still be a matter of some scientific uncertainty (at least until the point the damage was experienced).

It is difficult to underestimate the significance of the ambiguity of 'dangerous' climate change given that it is the ultimate objective of the convention. The problematic implications of climate change are, therefore, somewhat undefined, and there is no clear standard against which decisions made under the UNFCCC can be evaluated for their contribution. There is some scientific consensus (even if there is no political agreement) that the most serious consequences of climate change can be avoided if global average temperatures are limited to no more than 2°C above pre-industrial levels (Schellnhuber et al, 2006). From a small island developing state's (SIDS) perspective, however, even this level of warming is too high, as it means a commitment to sea-level rise and coral bleaching that in some locations may have catastrophic consequences for social-ecological systems.

That dangerous climate change remains poorly defined says much about the constraints to the UNFCCC's ability to solve the problem of climate change. Another major constraint is the absence of rules defining how decisions are to be made (a problem common to many UN forums). The default position, then, is that decisions are not possible without the consent of all parties and an objection to a decision by any party is a veto (at least initially). This gives any party the power to delay if not prevent decisions. In practice the objections of one party or small groups of parties eventually get overridden: yet it is not clear how many objections it takes to delay or obstruct a decision (Oberthür and Ott, 1999). So, because there is some value in the pluralism associated with the meaning of dangerous climate change, and because a sufficient number of parties do not wish to see an environmentally effective and binding climate regime (which would be more likely with a clearer definition of Article 2 that stipulates unacceptable levels of concentrations of greenhouse gases), attempts by scientists to sharpen the definition of dangerous climate change within the UNFCCC have failed. It also says something about the influence of governments in the IPCC process that despite some degree of scientific consensus that warming of 2°C above pre-industrial levels would constitute dangerous climate change, the IPCC assessment reports do not refer to this level of warming as 'dangerous'.

The convention established a conference of the parties to meet regularly and evaluate progress (Article 7). Thus since 1995 there has been an annual Conference of Parties (COP) to the UNFCCC. The COP is the most important, well-attended and widely reported of the meetings associated with the UNFCCC. However there are many others, including the meetings of its two main subsidiary bodies: the Subsidiary Body for Scientific and Technological Advice (SBSTA), and the Subsidiary Body for Implementation (SBI), both of which meet twice a year (once concurrently with the COP). There are also smaller working groups, technical workshops, committees and boards associated with the UNFCCC process. Keeping abreast of the developments emerging from all these meetings is a challenge for even the developed countries, and for most SIDS it is impossible.

For the SIDS, Article 4.8 of UNFCCC, which observes that full consideration should be given to meeting the needs of small island countries to respond to the adverse effects of global warming, is very important as it implies that all countries have a commitment to assist them with adaptation. The convention also commits developed country parties to help developing countries that are particularly vulnerable (which includes small island states) to meet the costs of adaptation (4.4). However, there is no specific indication of what form this assistance should take, and how it is to be achieved. Consequently, the SIDS have directed a considerable amount of their energy to seeking some progress on the implementation of these articles (and their related articles in the Kyoto Protocol). This progress has been impeded by some developing countries – in particular by the Organization of Petroleum Exporting Countries (OPEC) – as much as it has been by the developed countries (Barnett and Dessai, 2002; Barnett, 2008b).

Article 4 of the convention sets out commitments of the signatories, the first of which is to report on emissions and measures taken to reduce them (Article 4.1). Article 4.2 commits developed countries (identified as Annex 1 countries) to 'adopt national policies and take corresponding measures on the mitigation of climate change, by limiting [their] anthropogenic emissions of greenhouse gases and protecting and enhancing [their] greenhouse gas sinks and reservoirs' (Article 4.2a). However, these commitments are not binding for the signatories and no specific targets for reduction were included in the UNFCCC, although at the time of signing many developed countries had committed to emissions reductions targets ranging from stabilization at 1990 levels to cuts amounting to 25 per cent below 1990 levels by the year 2005 (Paterson, 1996).

In the five years after it was signed the most important outcomes of the UNFCCC were that the issue of climate change was brought to international political prominence, and that parties were committed to reviewing their progress towards reduction of greenhouse gas emissions (Grubb et al, 1999). As the Second Assessment Report of the IPCC, in 1995, provided stronger evidence of global warming, and it was becoming clear that reducing emissions to 1990 levels would not be enough to ameliorate climate change, there was a push for more significant cuts to be achieved under a binding protocol to the UNFCCC. The Alliance of Small Island States (AOSIS) played a significant role in this, pushing for 20 per cent reduction below 1990 levels by 2005, a figure that was picked up by many international environmental NGOs. While the reductions that were eventually agreed to were much less, the actions of AOSIS ensured that the issue of reduced emissions remained central to the negotiations thereafter (Grubb et al, 1999).

At the third conference of the parties (COP3) to the UNFCCC, in Kyoto in 1997, a more specific set of commitments to reduce greenhouse gas emissions was agreed upon, and became formalized in the Kyoto Protocol. The Kyoto Protocol sets relatively modest but legally binding targets for emissions reductions or limitations on 39 developed and 'economies-in-transition' countries listed in its Annex B (which includes most of the same countries listed in Annex 1 of the UNFCCC). The principle of 'common but differentiated responsibility' was applied in negotiations over these targets. The principle holds that while all countries are responsible for action on climate change, the developed countries should be the first to take on emissions reductions targets because they are responsible for most of the historical and current emissions of gases, are more able to reduce emissions (in terms of potential costs and in terms of the size of gains relative to present emissions profiles), and because people in developing countries still emit few greenhouses gases and should have some right to emit more as they pursue pathways out of poverty.

Australia, Russia and the United States long refused to ratify the protocol. Russia acceded to the protocol in November 2004, and 90 days later the protocol entered into force. It was only after a change in government in late 2007 that Australia ratified, making for uncomfortable relations between Australia and its Pacific Island neighbours for a decade. Of all the countries listed in Annex B only the United States is yet to ratify. In total 184 countries have signed and ratified the Kyoto Protocol. While the protocol is a necessary first step, the reductions it will achieve by 2012 will do almost nothing to slow the rate of climate change given that a 75 per cent reduction below current levels of emissions is, arguably, necessary (Stern, 2007). The first 'commitment period' for the Kyoto Protocol (the date agreed reductions have to be achieved by) is from 2008 to 2012. For 2012 a new set of targets is to be finalized (at COP15 in Copenhagen in 2009).

Within the Kyoto Protocol there is an instrument that enables countries listed in Annex B to the Kyoto Protocol (and which have ratified it) to meet some of their commitments through activities that reduce emissions in developing countries. The instrument – called the Clean Development Mechanism (CDM) – in theory allows for commitments to be achieved at the lowest cost, on the assumption that it is cheaper to reduce emissions in developing rather than developed countries. However the CDM is contentious, and its contribution to reducing emissions and promoting sustainable development is questionable (Olsen, 2007).

Because their emissions are so small, the SIDS have never been likely places for CDM projects, although the larger countries such as those in Melanesia could be engaged in projects that seek to sequester carbon through reforestation. Nevertheless the SIDS have an interest in the CDM for two reasons. First, the SIDS seek an environmentally effective climate change regime, whose activities lead to genuine reductions in emissions of greenhouse gases in a manner that is sustainable and equitable. It was never obvious that this was something that the CDM would achieve, and so the SIDS have monitored it closely. Second, 2 per cent of the 'certified emissions reductions' generated by CDM projects are allocated to an Adaptation Fund to fund 'concrete' adaptation projects and programmes in countries that are particularly vulnerable to the adverse effects of climate change (UNFCCC, Decisions, 10/CP.7 and 17/CP.7). After intense negotiations between 2001 and 2007 about the nature and governance of the Adaptation Fund, at COP13 in Bali the Adaptation Fund Board was finally established (Adaptation Fund Board, 2008). The policies and guidelines for the Adaptation Fund Board are provisional; however, they do include a list of sectors that would be eligible for funding (see Table 5.1).

The implication of the notion of 'concrete' projects, borne out by the sectors and activities in Table 5.1, is that projects should focus on material changes that are likely to be in the form of technical and engineering responses. This reflects a desire from the PICs, and other countries, that assistance for adaptation moves beyond what has thus far been piecemeal funding for activities that have largely been of an 'enabling kind': short-term projects funded by the GEF and bilateral donors that seek to collect information, build 'capacity', and produce reports, policies and plans. After a decade of projects such as these, and after many documents that list adaptation needs and priorities – such as the Initial National Communications to the UNFCCC and the National Adaptation Plans of Actions – many PICs feel frustrated that donors continue to fund only enabling activities. This is perceived as donors avoiding their responsibilities under the UNFCCC.

Table 5.1 *Sectors and activities eligible for Adaptation Fund support*

Sectors	Specific activities within sectors
Risk management (alert systems, prevention, insurance)	Activities supporting concrete adaptation technology transfer.
Agriculture	
Dry land management	Infrastructure and civil works.
Water resources (including water infrastructures)	
Health	Skill training, workshops and conferences.
Coastal zones (including integrated coastal zone management)	
Infrastructure development (roads, habitat, urban planning)	Capacity building, vulnerability assessment (only if necessary in order to implement project).
Fragile ecosystems (including mountain forest ecosystems)	
Biodiversity	
Forests	
Wetland management	

Source: Adaptation Fund Board, 2008, p8

The frustration is exacerbated by some negative experiences that the PICs have had with the GEF, including their observation that large amounts of GEF funding end up going to consultants rather than being spent within their countries, and the repeated participation of personnel in workshops and training meetings to improve their 'capacity' to do things that they often already know how to do or do not need to do, and which implies that they do not have 'capacity' in the first place.

Yet there remain some serious tensions between the desire in PICs for concrete projects and their need and effectiveness. There are three good reasons why programmes that build adaptive capacity remain necessary, particularly at the community level. First, there are concerns about the timing and magnitude of climate impacts, as well as about how effective some adaptation responses may be in reducing vulnerability. There is the possibility that investments in solutions to meet impacts whose timing and magnitude is uncertain, and whose effectiveness is uncertain, may result in high-cost material interventions that may be regretted (Barnett, 2001). Second, adaptation is a process in which communities will be required to make decisions at different times to adjust (proactively where possible) to changing environmental conditions. One-off projects are not likely to be successful in this regard, and indeed, if climate and consequential environmental changes carry on unimpeded, or only partly mitigated, these projects are likely to become outdated and ineffective. Third, capital-intensive adaptation projects may have adverse social and environmental effects (that is, they may create maladaptations). As is the case with all aid projects, minimizing the risks of adverse social and environmental outcomes from adaptation projects requires that they be guided by principles of effectiveness, fairness and transparency, and seek decision-making processes that are participatory and inclusive of all stakeholders (Barnett, 2008a). Yet there is very little in what donors or PICs say about more concrete adaptation projects to suggest that these issues of principle and process have been given much thought.

For these reasons, adaptation programmes should aim to develop a society's capacity to cope with change by building up its institutional structures and human resources, while maintaining and enhancing the integrity of its ecosystems, and while developing and refining processes for implementing projects at the community level. This does not mean that where 'concrete' adaptation projects are identifiable and properly executed they should not be funded, nor does it mean that the endless 'enabling' activities that frustrate the PICs should be continued. What we are saying though is that a rush to build things may create more problems than it solves, and should not come at the cost of community-level initiatives to build generic adaptive capacity. Both concrete and adaptive capacity-building initiatives are necessary, and both should be funded as part of developed-country commitments under Articles 4.8 and 4.9 of the UNFCCC.

The problem here is perhaps one of terminology: the concept of 'capacity' too often means capacity to comply with the processes of climate change institutions such as the UNFCCC and the GEF rather than capacity to adapt to climate change *per se*; and 'enabling' has meant reporting information that donors need rather than empowering communities to adapt to climate and other stressors. Throughout the discourses about climate change in PICs that circulate in the UNFCCC negotiations and GEF documentation, the phrase 'lack capacity' is frequently used, but rarely with reference to the task that needs to be achieved, or the financial, institutional as well as human resources required to achieve it. Thus 'capacity building' erroneously infers that 'people in the region are not able to do much'. Discussions about capacity must be more specific than this lest they be merely pejorative statements about the perceived inadequacy of others. Such discussions must consider the specific task to be performed (capacity to do *what?*), and the barriers that may impede the completion of the task, which are far more than the supposed deficiency of people, and can include financial constraints, information deficiencies, problems with institutional alignment, and perceptions of the need for and value of the task to be performed.

Participating in the UNFCCC/Kyoto Protocol processes places a considerable demand on Pacific and other small countries. At the Kyoto Conference of the Parties nearly 10,000 people were involved (Grubb et al, 1999), and at COP13 in Bali, 2007, almost 11,000 people participated in the meetings and associated side events, including 3500 government officials, 5800 representing multilateral, intergovernmental and non-governmental organizations, and 1500 accredited media representatives. As Table 5.2 shows, the Pacific Islands struggle in terms of the number of people representing them at the climate change negotiations: at COP14 the 14 independent nations had a total of 100 delegates (ranging from 25 from Papua New Guinea to two from Fiji) between them, which is very few given that there were 3958 delegates in total, averaging 21 delegates from each country. Delegations from the Pacific had on average seven members, and if we exclude Papua New Guinea the average drops to 5.8. On the other hand those island societies that are dependent territories of other states – such as French Polynesia, New Caledonia, American Samoa, Northern Mariana Islands and Tokelau – have no representation, and are dependent on their colonial powers to represent their interests. Given that the United States, a colonial power, even refuses to ratify the first set of greenhouse gas reductions and New Zealand is moving slowly, their interests are scarcely being represented at all.

Table 5.2 *Comparative data on delegation size at COP14, Poznan*

Country	Number of registered participants
Cook Islands	7
Federated States of Micronesia	8
Fiji Islands	2
Kiribati	3
Marshall Islands	6
Nauru	3
Niue	3
Palau	7
Papua New Guinea	25
Samoa[a]	14
Solomon Islands	5
Tonga	4
Tuvalu	8
Vanuatu	3

[a] This includes four participants from the Secretariat of the Pacific Regional Environment Programme which has its headquarters in Samoa.

Note: This table lists those who were registered participants. Not all of these would have actually attended the meetings and some countries may have had additional members on their teams.

Source: UNFCCC, 2008

The small size of the Pacific delegations creates considerable pressures for the members of delegations, who have very few people to cover a very wide range of issues being negotiated in parallel sessions. It is also the case that the Pacific countries typically send climate change officers to the negotiations, most of whom have backgrounds in meteorology or environmental science, and who rarely have training in international law. In contrast, the developed and larger developing countries send teams of international lawyers and other staff well-versed in international negotiating processes. The complexity of the issues and the nature of the negotiating process make it almost impossible for a Pacific country to single-handedly keep abreast of any but a couple of issues being negotiated at the COPs. For the countries that send even these few delegates, the costs of participation are high, often amounting to up to four weeks absence of some of their most skilled staff, of which there are typically very few to spare.

Nevertheless, the Pacific countries are a notable presence at the COPs, and have a disproportionate influence in the negotiations. There are three reasons for this. First, there is now a cadre of individuals from the Pacific who have a considerable amount of experience with the UNFCCC process, and a number of these are confident enough to actively engage in contact groups and have an influence in their outcomes. Second, the PICs receive support in the negotiations from the Secretariat of the Pacific Regional Environment Programme (SPREP), although the number of people SPREP sends is also small (four delegates at COP14). Third, the Pacific SIDS have formed an Alliance of Small Island States (AOSIS) with other SIDS from around the world, which generally acts to coordinate the participation of all SIDS in the negotiations on issues of common interest.

AOSIS has been effectively supported by the Foundation for International Environmental Law and Development (FIELD), an NGO which has assisted small island states at international climate negotiations since the Second World Climate Conference, and which has been a major collaborator with AOSIS since that time. FIELD takes a significant role in campaigning for substantial emissions reductions and also assists small island states at the various international meetings on climate change. FIELD is a subsidiary of the International Institute for Environment and Development (IIED), which publishes *Tiempo*, a quarterly magazine dedicated to climate change issues in developing countries and which has given considerable coverage to Pacific Island issues. FIELD has probably been the most significant international environmental NGO supporting SIDS, including those from the Pacific region (the role of AOSIS is discussed in greater detail later in this chapter).

Positions in the negotiations over the UNFCCC and the Kyoto Protocol are often said to be polarized between the North (the wealthy OECD countries) and the South (the remaining developing countries). However, this is a crude characterization of what is a far more complex and dynamic political landscape. Most developing countries agree that the great majority of anthropogenic greenhouse gases in the atmosphere are the result of emissions from the OECD countries over the past 250 years. They argue that a uniform reduction in emissions applied to all countries would deny them the chance to industrialize and, in so doing, to reach the same levels of development that the North has enjoyed. This stance is particularly strong among the three newly developing giants of Brazil, India and China (the BrIC), and has considerable logic. But for small island states the implication of this right to emit is unacceptable, as their rights to develop and possibly exist will be significantly curtailed. For example, during COP13 the Vice President of the Federated States of Micronesia, Alik Alik, observed that by procrastinating, countries at the meeting were in effect making a far reaching moral decision – that some people and cultures may be expendable:

> *By delaying actions relating to climate change, the world should not be*
> *saying that the lives of some people are more important than those of*
> *others. Nor should it imply that some cultures are more worthy of saving*
> *than others. (Islands Business, 12 December 2008)*

There are other areas of tension among members of the South. For example, OPEC claims that efforts to control emissions may undermine their income from exports of oil, and so tends to oppose any policy or measure that may reduce demand for oil (Barnett et al, 2004). This is directly at odds with the concerns of SIDS which are pressing for significant reductions in greenhouse gas emissions. Thus it is inaccurate to say that all developing countries see the issues in the UNFCCC in the same way.

Developed countries also have diverse goals. The United States, for example, cannot yet manage to ratify the Kyoto Protocol, whereas the EU is arguing for further emissions reductions beyond those agreed to in the Kyoto Protocol. Indeed, the difference between these latter two is so great that the EU has been described as a 'protagonist' to the United States on the issue of climate change (Vogler and Bretherton, 2006).

At the heart of many tensions in the UNFCCC lies the recognition that climate change mitigation will require significant changes in the idea and practice of development itself. The causes of climate change are most often characterized as being proximate processes such as car exhaust systems, factory chimneys, rice paddy production, and the chainsaws and torches of deforesters. Nearly all of the focus on mitigating climate change has been on technical and technological solutions to these proximate causes – what Mol (2001) calls an ecological modernization response. This reflects the dominant liberal discourse that seeks to maintain current models of economic growth.

However, it is unlikely that such measures will be sufficient to reduce greenhouse gas emissions. As Mr Pateson Oti, the Solomon Islands Minister of Foreign Affairs told a press conference prior to the Commonwealth Foreign Ministers Conference in 2007, on behalf of AOSIS: '[c]limate change is the symptom and not the disease. The disease is our unsustainable means of production, worsened by unsustainable patterns of consumption' (Holdsworth, 2007). The larger developing countries are aspiring to achieve, and are fast nearing, patterns of consumption similar to those in the developed world, and they, like the developed countries, feel that development as they perceive and desire it is at risk (from climate change mitigation). Of course, most of the individual climate negotiators from the large developing countries are as wealthy, and probably contribute similar volumes of emissions, as their developed country counterparts.

Unfortunately, it is not clear how the consumption disease can be cured. Ecological modernists see opportunities for green consumerism, and many governments and environmental groups are encouraging their citizens to consider green products. However, the United Nations Environment Programme

(UNEP), in a recent publication promoting critical consumerism as a means of reducing individual carbon footprints concludes that reduction of consumption is necessary, and that:

> Going on a climate diet will not be exactly fun, either, though it may help us to rediscover the forgotten delights that come from doing more with less. But it will give us and future generations the hope of survival on a sustaining Earth. (UNEP, 2008, p15)

There has been very little effort given to looking at other economic models in which rampant consumption is minimized. It may be that until such a model is found and applied, the underlying causes of climate change are likely to prove intractable. This is perhaps the true challenge – one that the UNFCCC thus far seems wholly incapable of addressing.

The Global Environment Facility

One issue on which most of the developing countries tend to agree (and on which most of the developed countries tend to disagree) is that the main financial instrument of the UNFCCC – the Global Environmental Facility (GEF) – does not serve the interests of developing countries well.

The GEF was established in 1991 by the United Nations Development Programme (UNDP), UNEP and the World Bank, to promote sustainable development projects in developing countries. It was identified in the UNFCCC as an interim financial mechanism for implementing the Convention. The GEF was restructured in 1994, giving it greater autonomy from the World Bank, and it is now a financial mechanism for implementing a number of international environmental treaties including those on biodiversity, international waters, desertification and persistent organic pollutants, as well as climate change. Between 1992 and 2005, the GEF made available US$4.5 billion in grants, supplemented by US$14.5 billion in additional funding, for 1400 projects in these areas (GEF, 2005). Of this, about a third was allocated to climate change, almost all of which was for large-scale mitigation projects – over half concentrated in ten large industrializing countries, most notably China (Mace, 2005). Of the US$4.5 billion, around US$83 million was allocated to climate-change related projects in SIDS, and of this less than 10 per cent (a little less than US$7 million) was allocated to PICs. As Mace argues, the GEF's

> near-exclusive focus on mitigation activities has led to frustration for many SIDS and other countries vulnerable to the impacts of climate change, who have called repeatedly during the climate negotiations for greater attention to convention commitments on adaptation. (Mace, 2005, p228)

The majority of the money spent in PICs has been for enabling activities to assist PICs to meet their reporting requirements under the UNFCCC. About US$1.4 million was allocated for renewable energy projects, and until 2005 nothing specifically had been allocated for adaptation, although the reporting required assessments of vulnerability and adaptation. Since the 2005 report on GEF activities in relation to SIDS, a major renewable energy project has been endorsed (GEF funding US$5.225 million; co-financing US$20.8 million) for countries in the Pacific Islands region.

The GEF reports that by 2007 it had made available US$7.4 billion in funding, with over US$28 billion in additional funds, for almost 2000 projects in addition to over 7000 small grants to communities and NGOs. Of the US$7.4billion, US$280 million had been made available for adaptation through three funding mechanisms: the Strategic Priority for Adaptation (SPA); the Least Developed Countries Fund (LDCF); and Special Climate Change Fund (SCCF) (GEF, 2007). This figure represents roughly 3.8 per cent of all GEF funding that has been allocated, and reflects the slow progress made in supporting adaptation. The main aim of these funds is to 'mainstream' adaptation into sustainable development activities – mainstreaming in this sense largely means taking into account climate risks when making decisions.

These funds have contributed to two national projects in the Pacific region thus far: US$1.8 million for the Kiribati Adaptation Pilot Project (under the SPA), and co-financing of US$2 million for an Integrated Climate Change Adaptation project in Samoa (under the LDCF). The LDCF also funded the preparation of National Adaptation Plans of Action (NAPAs) in Kiribati, the Solomon Islands, Vanuatu, Samoa and Tuvalu. The NAPAs were to report on urgent requirements for adaptation. Table 5.3 summarizes the reports and indicates that all of the countries identified agriculture/food security and water resources as among their priorities for adaptation projects. The costs of these projects should not be read as indicating the total costs of adaptation in these countries (Agrawala and Fankhauser, 2008).

In total the GEF distributed US$5.8 million for adaptation in the Pacific between 1994 and 2008, almost all of which was for enabling activities (McGoldrick, 2009). This is remarkably little given the long-term recognition of the particularly high vulnerability of the PICs, the recognition of the need to provide assistance for adaptation in the UNFCCC and the Kyoto Protocol, and the much repeated and insistent demands from PICs that adaptation be a priority area of activity for the GEF. Even this modicum of funding was hard won, and accessing it was extremely difficult, for the reasons outlined below. It was not surprising then, that the Pacific SIDS, along with most other SIDS and many of the LDCs and other developing countries, were very reluctant to have the GEF be the institution responsible for administering the Adaptation Fund.

Table 5.3 *Priority adaptation projects submitted in NAPAs*

	Kiribati US$ million	Samoa US$ million	Tuvalu US$ million	Vanuatu US$ million	Solomon Islands US$ million
Agriculture/food security	0.450	0.320	2.220	1.000	6.500
CC info/monitoring/upgrading met services/early warning systems	0.377	4.500			
Conservation		0.350			
CZM	6.415	0.450	1.907		1.750
Disaster risk reduction			0.462		
Enabling effective participation at international forums	0.060				
Fisheries/marine resources/coral					1.500
restoration	0.499		1.025	1.000	
Health		0.620	0.382		
Infrastructure development					2.000
Land use management		0.400			
Low-lying and artifically built islands					3.500
Project management/institutional strengthening	0.234				
Reforestation/sustainable forestry		0.418		1.000	
Sustainable tourism		0.250		1.000	
Tourism					0.500
Waste management					1.500
Water resources	2.321	0.505	2.675	1.000	
Total estimated costs of priority adaptation projects	10.356	7.813	8.671	5.000	17.250

In 2008 a large Pacific regional project was approved with funding of more than US$13 million (under the SCCF), and with co-financing of almost US$40 million (nearly all from the governments of the countries where the projects are to be implemented). This Pacific Adaptation to Climate Change project (PACC) focuses on three sectors – food security, water supply and coastal management – and seeks to develop national adaptive capacity (through developing technical capacity, institutional coordination, tools and legislative and policy directives). There will also be community-level activities on climate change risk-management in agriculture and resilient water management (GEF, 2008).

Despite these activities small island states, including those from the Pacific, have found it difficult to access funds from the GEF for adaptation projects. There are several reasons for this. First, the various GEF funds require consider-able bureaucratic input which makes them difficult for small countries to access without considerable cost and reliance on consultants. This creates the potential for consultants to establish the agenda for climate change adaptation, based on their disciplinary and individual preferences. Countries also rely on regional agen-cies such as SPREP to access GEF funding. In turn, SPREP must work with one of the few agencies able to implement GEF funds, which in the Pacific is most often UNDP, but has also included UNEP and the World Bank. In most cases the implementing agencies and regional agencies take a share of the funds allocated by the GEF to projects to cover administrative costs. Implementing GEF projects becomes a business activity for implementing agencies and regional organizations, and it is usually in their interests to keep securing GEF projects. This creates some interesting dynamics between countries and agencies, and between agencies and the GEF.

In addition, adaptation funds are not just for Pacific, or indeed SIDS in general, but available to a wide range of developing countries that will need to adapt (Nurse and Moore, 2005) and the quantum of financial resources indicated thus far is inadequate for even a modicum of the costs of adaptation. Moreover, the LDC fund excludes the majority of PICs. For a long time the GEF required that a project yield 'global environmental benefits', where these were defined by the treaty under which funding was allocated (Mace, 2005). In practice this made it difficult to argue that, say, implementing a dengue fever monitoring system was something that would provide a 'global environment benefit', even if it was a response to a global environmental risk. Local environmental benefits have therefore been seen by the GEF to be distinct from global ones, and not eligible for funding (Mace, 2005). But perhaps the most problematic requirement has been that the GEF will only provide funding for the additional costs of actions that arise because of climate change. This remains a poorly defined requirement, but implies that 'baseline' costs associated with developments that governments would fund anyway must be met by governments, and the GEF will only fund the extra costs associated with ensuring those developments are sustainable in the face of climate risks. Some of these criteria are supposed to have been relaxed in the case of projects implemented under the SCCF, the LDCF and the Adaptation Fund. However, this is yet to be seriously tested, and the Pacific SIDS remain wary of the GEF and its obscure and difficult bureaucratic processes.

Adaptation, and the GEF's role in financing it, remains a bone of contention for PICs and other SIDS. COP14 broke up on a relatively rancorous note as Annex 1 parties failed to agree to provide the kind of financial resources that would be required for meaningful adaptation, and many of the conditions imposed by the GEF do not seem to have been removed. This frustrated a number of Pacific Island delegations, including the Prime Minister of Tuvalu who was greeted with applause from the audience at Poznan when he observed:

> But I have to say that I am deeply worried by the way negotiations on the Adaptation Fund are progressing at this meeting. SIDS like Tuvalu need direct access and expeditious disbursement of funding for real adaptation, urgently. We are suffering already from effects of climate change.
>
> I am compelled to say that we are deeply disappointed with the manner in which some of our partners are burying us in red tape. This is totally unacceptable.

> *We don't want the Adaptation Fund to turn into other funds administered by the Global Environment Facility, where only countries that can properly access this fund are the ones that can afford consultants and UN agencies to write lengthy and endless project proposals and work their way through metres of red tape and survive lengthy delays. (Reported by Makereta Komai, Islands Business, 2008)*

While the UNFCCC and Kyoto Protocol fail to achieve anywhere near the mitigation requirements to ensure the safety of many island countries, they have also failed to provide the assistance to help them cope that was first signalled in 1992.

The Alliance of Small Island States (AOSIS)

In the late 1980s it became apparent to those people from SIDS who were negotiating and campaigning in international forums for an international response to climate change that the task was much greater than could be managed by each country working independently. It was very early in the international climate change process – in 1990 at the Second World Climate Conference where Bikenibeu Paeniu, Prime Minister of Tuvalu made his speech – that the Alliance of Small Island States was formed. AOSIS is a coalition of countries that share a common concern for the environment. AOSIS's work as a coalition means that the interests of countries are advanced more than they would be if they each acted independently. AOSIS, for example, was one of the few developing-country groups to play an active role in the development of the UNFCCC (Shibuya, 1996). In the case of climate change, it might also be asserted that AOSIS perhaps wields greater moral authority than most if not all other countries, with some considering it to be the conscience of the climate change negotiations (Davis, 1996).

AOSIS has 39 member states and four observers which are island territories of metropolitan countries (see Table 5.4). Four of AOSIS's members are small non-island states on the coastal margins of continental land masses. There are tensions between those AOSIS members who are members of G77 or have aligned themselves with the G77, which holds the view that emissions reductions should be restricted to the Annex 1 parties, and those that are not part of the G77 (such as the Cook Islands, Niue and Tuvalu) or which hold the view that the large developing countries must also begin reducing their emissions. As Nurse and Moore (2005) point out, those who align with the G77 view are completely at odds with AOSIS proposals for significant reductions in emissions globally, and they suggest that some conditions could be usefully placed on membership of AOSIS. This point was also raised in 2004, when Palau argued that there is a growing gap between the needs of G77 and its AOSIS members (Note et al, 2004) and suggested the need for an AOSIS charter.

Table 5.4 *AOSIS member states*

Caribbean Island states	Pacific Island states
Antigua and Barbuda	Cook Islands
Bahamas	Fiji
Barbados	Federated States of Micronesia
Cuba	Kiribati
Dominica	Marshall Islands
Dominican Republic	Nauru
Grenada	Niue
Haiti	Palau
Jamaica	Papua New Guinea
St. Kitts and Nevis	Samoa
St. Lucia	Solomon Islands
St. Vincent and the Grenadines	Tonga
Trinidad and Tobago	Tuvalu
	Vanuatu
Indian Ocean island states	**Atlantic (African) Island states**
Comoros	Cape Verde
Maldives	Sao Tome and Principe
Mauritius	
Singapore	**Non-Island states**
Seychelles	
	Belize
Mediterranean Island states	Guinea-Bissau
	Guyana
Cyprus	Suriname

There are other tensions too, such as those between Singapore, which is wealthier in per capita terms than some OECD countries, and which is seeking to avoid an emissions reductions target under a successor agreement to the Kyoto Protocol, and other SIDS who argue that wealthy countries such as Singapore should reduce their emissions. Within the Pacific SIDS there are also tensions between the larger countries, in particular Papua New Guinea, that are seeking to earn certified emissions reductions for reducing emissions by controlling deforestation, and those who worry that such a scheme is unlikely to lead to any genuine reduction in emissions globally.

The Pacific Islands Forum

Having discussed the global policy framework in which SIDS act to pursue their interests on climate change, this section describes the role that the region's predominant regional body – the Pacific Islands Forum – has played in the politics of climate change in the region.

The Pacific Islands Forum is the annual meeting of heads of state and governments from the independent PICs, together with Australia and New Zealand. Table 5.5 lists the Forum member countries. The forum is supported by a secretariat which has had several name changes in the past few decades, but which is now called the Pacific Islands Forum Secretariat (PIFS). The secretary general of the PIFS is also chair of the Council of Regional Organisations in the Pacific (CROP) which coordinates the 11 major regional organisations (discussed further in Chapter 6).

Table 5.5 *Pacific Islands Forum member states*

Members
Australia
Cook Islands
Federated States of Micronesia
Fiji (suspended since 2 May 2009)
Kiribati
Marshall Islands
Nauru
New Zealand
Niue
Palau
Papua New Guinea
Samoa
Solomon Islands
Tonga
Tuvalu
Vanuatu
Associate members
French Polynesia
New Caledonia
Observers
Asian Development Bank
Commonwealth Secretariat
Timor-Leste
Tokelau
Wallis and Futuna

There have been 39 meetings of the Pacific Islands Forum since the first was held in 1971 (attended by five PICs and Australia and New Zealand). There are no formal rules governing the conduct of forum meetings, and so decisions are reached by consensus. The topics discussed at the forum are a useful barometer of the salient issues confronting the region: previous topics have included issues such as trade liberalization, regional transport networks, fisheries development, decolonization and security. The forum develops policy on regional approaches to issues such as crime, telecommunications and sustainable development. A major recent initiative of the forum has been the development of the Pacific Plan, a blueprint for sustainable development in the region. After each meeting a communiqué is issued which describes the issues that were discussed and the outcomes of discussions, provides an update on the progress of regional policies and plans that are being developed, and which sometimes includes the text of any relevant agreements, declarations and plans. The communiqué is the basis for regional policies and the work programme of the secretariat.

Environmental issues have been central to the forum's discussions. Figure 5.1 shows the major environmental issues included in forum communiqués over time. It shows that environmental issues were included in the very first forum, when French nuclear arms testing was of considerable concern to forum members. While in this first instance the opposition was to the actions of a single country (France), which was continuing with atmospheric nuclear weapons testing in French Polynesia, there were dimensions of this problem that were to resonate with later concerns about climate change: the impacts were likely to be felt throughout the region, and the cause of the problem was an outside power over whose actions the region had little influence.

Other environmental issues included concern about proposals by the United States to dispose of radioactive waste in the region, giving rise to The South Pacific Nuclear Free Zone Treaty, endorsed by the forum in Rarotonga in 1985. Fisheries issues have also featured in the communiqués from the first forum meeting, but have tended to relate more to obtaining benefits from the fisheries

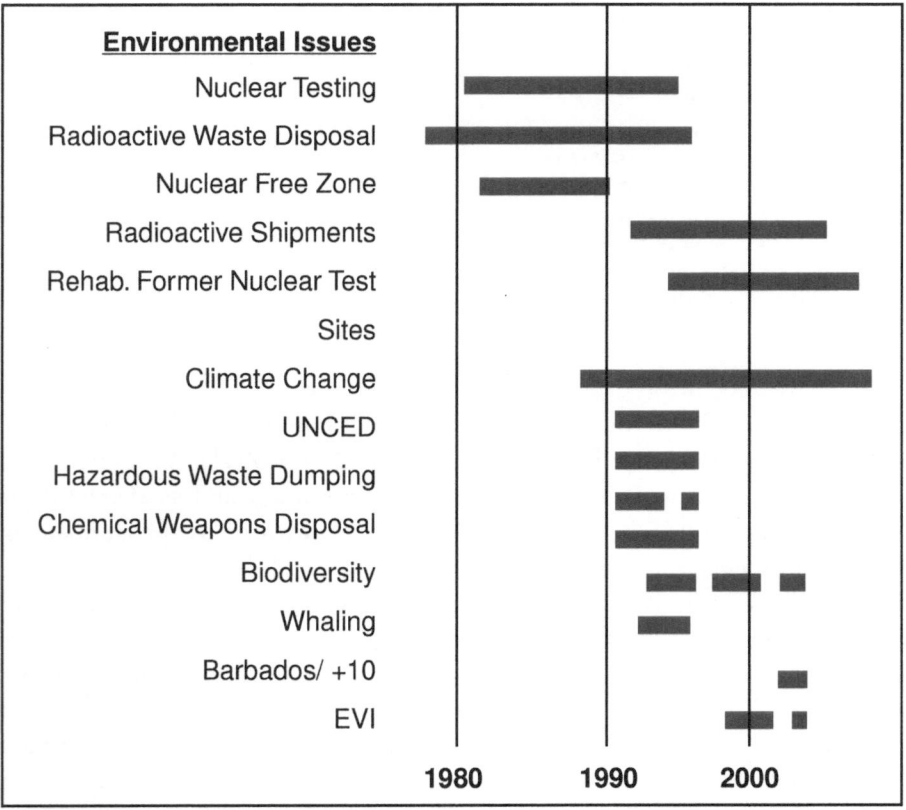

Figure 5.1 *Environmental issues discussed at Pacific Islands Forum meetings*

and stopping the exploitation of the resource by foreign vessels than issues of sustainability (although in 1989 the forum decided upon the Tarawa Declaration, calling for a ban on destructive drift net tuna fishing practices).

Climate change was first discussed at the forum in 1988, when it called for more information and for monitoring of its effects in the region. The issue has remained on the agenda thereafter and the level of concern expressed has increased over time, as have the forum's exhortations for industrialized nations to commit to reduced emissions of greenhouse gas emissions and, more latterly, for urgency in addressing both mitigation and adaptation issues. In 1990, 'Environment' was for the first time the chief focus of the forum.

The forum serves an important function in providing an opportunity for the leaders of independent PICs to meet and discuss the issue of climate change, and to exchange views with Australia and New Zealand on the issue. In 2005 the forum endorsed the Pacific Islands Framework for Action on Climate Change (PIFACC), which SPREP implemented in the form of a Pacific Islands Action Plan on Climate Change. A Pacific Climate Change Round Table has been established to meet yearly and oversee progress on the action plan and ensure that activities are coordinated. The framework and the plan have six sets of objectives to be achieved by 2015. These are:

1   Implementation of adaptation measures.
2   Governance and decision making (incorporating climate change into countries' sustainable development planning).
3   Improved understanding of climate change (focused on scientific and technical data collection and storage, modelling, and analytical frameworks with a brief reference to use of traditional risk management knowledge).
4   Education, training and awareness (including capacity building and public awareness).
5   Contributing to global greenhouse gas reduction.
6   Strengthening existing and building new partnerships and cooperation.

(SPREP, 2005, 2006)

The framework and plan clearly set adaptation as a regional priority and, given the lack of progress towards achieving a meaningful international mitigation programme, this is a wise course of action. They also set out a series of actions that will see increasing self sufficiency in PICs in relation to a range of climate change issues. However, they also clearly indicate the need for additional funding and see the international funding mechanisms that have emerged and are emerging from the UNFCCC and Kyoto Protocol processes as key in this regard, in addition to other multilateral and bilateral funding arrangements.

Climate change was the leading theme of the thirty-ninth forum meeting in Niue, and the meeting adopted the Niue Declaration on Climate Change. The Niue Declaration (PIFS, 2008) succinctly captures the position of the forum members on climate change, stating that they are:

> **DEEPLY CONCERNED** *by the serious current impacts of and growing threat posed by climate change to the economic, social, cultural and environmental well-being and security of Pacific Island countries; and that current and anticipated changes in the Pacific climate, coupled with the region's vulnerability, are expected to exacerbate existing challenges and lead to significant impacts on Pacific countries' environments, their sustainable development and future survival;*
> *and,*
> **RECALLING** *that despite being amongst the lowest contributors to factors causing climate change, the Pacific Islands region is one of the most vulnerable to the impacts of climate change including its exacerbation of climate variability, sea level rise and extreme weather events;*
> *and,*
> **RECOGNISING** *the importance of retaining the Pacific's social and cultural identity, and the desire of Pacific peoples to continue to live in their own countries, where possible;*
> *and,*
> *stressing the need for urgent action by the world's major greenhouse gas emitting countries to set targets and make commitments to significantly reduce their emissions, and to support the most vulnerable countries to adapt to and address the impacts of climate change;*
> **ENCOURAGE** *the Pacific's Development Partners to increase their technical and financial support for climate change action on adaptation,*
> *and*
> **CALL ON** *international partners to assist our development by undertaking immediate and effective measures to reduce emissions, use cleaner fuels, and increase use of renewable energy sources.*

That such a declaration could be endorsed by both Australia and New Zealand as well as the other forum members reflects the relatively supportive position these countries had towards the region on the issue of climate change at that time. However, as the next section shows, this has not always been the case, and cannot be guaranteed to be the case in the future. As climate change becomes an increasingly important problem, there is a risk that the policy goals of Australia and New Zealand on the one hand, and the PIC forum members on the other, may become so divergent that the inclusion of the two developed countries as forum members becomes increasingly untenable.

Australia and New Zealand

From the perspective of the PICs two important metropolitan actors in the region are Australia and New Zealand. They are both members of the Pacific Islands

Forum and signatories to the SPREP Convention. They are both significant aid donors, important trading partners, central to most transport routes, and an important source of education and health services. There is close cooperation on defence and policing between Australia, New Zealand and PICs. Further, there are large populations of Pacific Islanders living in both countries, and some countries share currencies with either Australia or New Zealand. In many ways then, the PICs' relationship with Australia and New Zealand is important for their sustainable development and stability.

Until recently, however, Australia and New Zealand have taken quite different trajectories from the PICs in terms of their commitments to the Kyoto Protocol. In the early 1990s, including during the first IPCC assessment, Australia and New Zealand, together with the Netherlands and the United States, co-chaired the Coastal Zone Management Subgroup of the Response Strategies Working Group, and they developed a close working relationship with many of the PICs. In the lead-up to the Earth Summit, both Australia and New Zealand pledged to cut emissions by 20 per cent below 1990 levels by the year 2000. But, as the political problems associated with reducing greenhouse gas emissions became more evident throughout the 1990s, Australia's earlier pro-climate change stance began to significantly weaken, so that by the late 1990s it was refusing to ratify the Kyoto Protocol despite its very liberal target of limiting emissions to 8 per cent *above* 1990 levels.

The years of the Howard government in Australia, from 1996 to 2007, were notable for the reluctance of the Australian government to take any action on climate change. There was even a reluctance to admit that climate change was a problem, and action taken was of the 'further research' kind (see Chapter 7). The Pacific Islands Forum meeting in Rarotonga in 1997 was a key event that revealed this disposition. This meeting occurred in the lead-up to the Kyoto COP, where the first protocol setting binding emissions targets was to be negotiated. At the forum the PICs were seeking a strong communiqué urging the developed countries to agree to deep cuts in emissions. The Australian Prime Minister refused to support such a communiqué, and argued that Australia should be allowed to increase its emissions by 18 per cent above 1990 levels (as opposed to the AOSIS proposal of a 20 per cent reduction). His stance and demeanour at the forum ruptured the meeting, with Tuvalu Prime Minister Bikenibeu Paeniu saying that 'Australia dominates us so much in this region ... we would have liked to have got some respect' (in Shibuya, 2004, p111), and Nauru President Kinza Clodumar questioning Australia's right to belong in the forum given its disregard for island issues (in Hussein, 1997; Fry, 1999). Howard's position seriously strained the consensus-based decision making in the forum, and used it to ensure the final communiqué contained a relatively uncontroversial statement about emissions targets. The Howard government later alienated a number of PICs with its heavy handed role in response to civil unrest in the Solomon Islands, and insistence that an Australian should head the forum secretariat.

Through the mid 1990s New Zealand reluctantly kept in step with international developments on climate change. It was part of the JUSCANZ (Japan, United States, Canada, Australia and New Zealand) group that wanted weak mechanisms for enforcing the protocol, and argued for the inclusion of flexible mechanisms to allow countries to meet the Kyoto commitments through activities in other countries (such as the CDM), and for special provisions related to Land Use, Land Use Change and Forestry. These demands were accommodated in the Kyoto Protocol in various ways, and they are generally seen as watering down its environmental effectiveness. The common interest that Australia and New Zealand had in CDM activities and Land Use Change resulted in clumsy attempts from both countries to attempt to lure the support of the larger Melanesian countries for their negotiation positions in the UNFCCC in exchange for promises of income from carbon sequestration and avoided deforestation activities. Had these attempts been successful, it would have broken PIC and also AOSIS solidarity on the issue of emissions targets and policies and measures to achieve them, an outcome that would not have upset either Australia or New Zealand.

For the most part, New Zealand's role in watering down the Kyoto Protocol went unnoticed in the Pacific Islands, and it certainly did not attract the ire of the region in the same way that Australia's more openly belligerent stance did. The fourth New Zealand Labour government, which held office from 1999 to 2008, ratified the protocol in 2002 and at least maintained the rhetoric of strong support for significant reductions, although its first commitment period target under the Kyoto Protocol of 100 per cent of 1990 emissions suggests that this commitment was somewhat less than other OECD countries. It had also committed to a domestic emissions trading scheme as a means by which its emissions could be reduced, but backed down on some other initiatives, such as a levy on pastoral farmers to finance research on reducing methane emissions from ruminants (which is a major part of New Zealand's emissions profile).

With the election of the Australian Labour Party in 2007, the new Prime Minister, Kevin Rudd, moved quickly on his election promise to ratify the Protocol, which he did at COP13 in Bali. One of Rudd's first acts after the election was to meet with his New Zealand counterpart to discuss climate change and the Bali meeting. This was greeted warmly by PICs, as was the Rudd government's assurance that its election promise of providing A$150 million to assist the region with adaptation would be kept. However, it was not long before this new climate rhetoric gave way. A year after ratifying the protocol, and during COP14, the new Prime Minister announced that Australia would be able to achieve no more than a 5 per cent reduction in the second commitment period (increasing to 15 per cent if other countries also did so), diminishing his lustre as a climate leader in the region. New Zealand, too, has shown colder feet under its newly elected right leaning National Party government, which has among other things announced that a select parliamentary committee will be established to examine the legitimacy of IPCC findings. Indeed, heading into the Copenhagen negotiations at the end of 2009, the climate policies of the two neighbours seem to be converging, and in a way that will offer little comfort to, and which will find little support in, the PICs.

Conclusions

PICs are affected by climate change in a number of ways. They can adapt to some extent by developing strategies to adjust to the effects of sea-level increases, warmer temperatures, changes to agro-climatic conditions, increases in incidences of vector and water-borne diseases, and reductions in freshwater quantity and/or quality. But, they can't develop a feedback mechanism on their own that effectively reduces the emissions that cause the problem, and if emissions are not very significantly reduced then there will be real and immediate limits to what adaptation can achieve. Therefore solving the problem ultimately relies on international cooperation which, as the progress of the UNFCCC indicates, is not easy.

Nor is progress in international action to address climate change something the Pacific Islands have much control over. Their power is constrained because they have small populations and could scarcely be more peripheral in the world economy. Further, despite their occupying 12 seats in the United Nations General Assembly, and 14 in the UNFCCC, their political efficacy is constrained because they cannot afford the money or personnel required to sustain large diplomatic missions and delegations to the UNFCCC. Indeed, as we have shown in Chapter 2, the processes of contact and colonialism, followed by independence and globalization have resulted in small formerly independent or interdependent communities in the Pacific becoming even smaller parts of a much greater system over which they have much less control than was previously the case – a process Brookfield (1980) called system enlargement.

Nevertheless, the SIDS do have some influence in the climate change negotiations because they are well organized, experienced, and well supported by FIELD. Most importantly though, they remain the key moral touchstones in the debates about mitigation and they remain the most compelling icon and send the clearest signal that reductions must be achieved. Yet thus far this has not proven to be sufficient moral suasion to encourage both the OECD countries and the rapidly industrializing countries of the South to make the kinds of reductions in emissions required to attenuate the climatic effects of greenhouse gas emissions. Indeed, some worry that islands serve the rest of the world as useful 'canaries' in the metaphorical mine of climate change, and it will be only when they succumb that there will be a sense of urgency among the rest of the world's nations (see Chapter 8 for a discussion of the canary discourse).

We are not optimistic that the climate change regime, or any other social order, will emerge that is capable of responding to climate change in a manner and in time to avoid significant, if not catastrophic impacts on PICs. Given this, the case for adaptation in PICs becomes even more pressing. This is not to say that adaptation is an alternative to mitigation; the two are not fungible, and if atmospheric concentrations of greenhouse gases continue to increase the costs of adaptation will grow and the limits to the ability of adaptation to avoid damages to the needs, rights and values of people in the Pacific will be reached (there will be adaptive failure). Rather, adaptation is the only response open to the PICs given the global failure to mitigate climate change.

For this reason it is doubly distressing that even as many greenhouse gas emitting nations procrastinate on efforts to reduce emissions, they are also extremely resistant to supporting wide ranging adaptation in PICs. As we have seen, contributions to the GEF have been woefully inadequate, and the GEF itself has been woefully incapable of supporting the PICs to implement adaptation. This places many SIDS, including PICs, in an invidious position. They do not make a significant contribution to global warming, are struggling to get support from even fellow developing countries for substantial commitments to reduce emissions, and are receiving precious little to help them face an uncertain future.

6

Doing Climate Change in the Pacific

The previous chapter suggested that for all that has been said about the vulnerability of the Pacific Islands to climate change, responses to the problem from the United Nations Framework Convention on Climate Change (UNFCCC) and the Global Environmental Facility (GEF) have been inadequate. For most of the past 20 years much the same can be said about the responses of other funding institutions. Until recently the multilateral development agencies and metropolitan countries that take an interest in the region have done little to assist with the task of adaptation, and mostly it is has been enabling type activities that have been supported. Many of the large NGOs that campaign on climate change by highlighting the particular vulnerability of Pacific Island countries (PICs) (see Chapter 8) have also done little in the way of applied responses in the region, although there have been two notable exceptions discussed later in this chapter. In more recent years, however, there has been a considerable increase in the number of projects being implemented in the region, largely because funding agencies are belatedly starting to respond to the oft-repeated calls from PICs for assistance for adaptation.

This chapter provides an overview of projects and programmes that have been implemented in the Pacific Islands in response to climate change. It focuses on projects that address adaptation, and examines in detail some of those that have offered positive lessons from which others could learn. These positive examples stand in contrast to the two somewhat more problematic and much more extensive projects discussed in the following chapter.

The PICs themselves are heavily engaged in activities within their own countries on climate change, which are funded entirely from within their own budgets. These include paying the salaries of public servants engaged in climate change work, arranging and hosting meetings and workshops of various kinds, funding various outreach and awareness-raising activities in communities, and including climate change themes in publicly funded media broadcasts. However, these activities are not well documented and reported relative to those funded by donors (which require extensive documentation), so they are not noticed by observers from outside, and there is little that we can say about them in this chapter. The same is also true for local NGOs engaged in climate change work: that their activities remain largely invisible to outsiders does not mean that they are passive actors in responding to climate change. An effort at documenting these endogenous responses is necessary, and we suspect would reveal that climate change is already being 'mainstreamed' into government activities in the region, albeit in a manner and at a pace that diverges from the expectations of donors.

This chapter does not discuss these endogenous responses to climate change in the region because we do not have enough experience with them to comment: they are not well documented; and they are best assessed by those people associated with them. We stress that these endogenous responses do exist and are significant although they are not recognized in this chapter or other writing. It cannot be said that the Pacific Islands simply use climate change as another cause for which they seek a rent from donors.

It is also important to recognize that the PICS commit considerable in-kind funding to most donor funded projects. This takes the form of the costs associated with the use of facilities, land and labour that would otherwise be used for other public good projects. In many cases the value of these in-kind commitments exceeds that of the cash provided by donors: for example the co-financing provided by countries to the recently implemented Pacific Adaptation to Climate Change (PACC) project is valued at close to three times the value of money provided by the GEF; and the co-financing of the GEF-funded Pacific Islands Greenhouse Gas Abatement through Renewable Energy Project (PIGGAREP) is worth nearly four times the value of the money provided by the GEF.

In this chapter we discuss responses in the region that we have had some experience with, and for which there is some documentation. The chapter begins with an overview of all projects in the region, noting that gaining such an overview is increasingly difficult as the number of projects multiply. We then provide an overview of the climate change activities of regional organizations in the Pacific, and discuss the first major regional project – the Pacific Islands Climate Change Assistance Programme (PICCAP). Next, we move from regional projects to those financed by single donors, and here we discuss in some detail the Capacity Building for the Development of Adaptation Measures in Pacific Island Countries (CBDAMPIC) project funded by the Canadian International Development Agency (CIDA), and the Climate Change Adaptation in Rural Communities of Fiji (CCA) project funded by AusAid. Finally, we discuss two projects funded by international NGOs in the region, the Climate Witness project funded by the World Wildlife Fund (WWF), and the Red Cross/Red Crescent Preparedness for Climate Change Programme (PCCP).

An overview of projects and programmes

In October 2008, 133 delegates from governments, regional organizations, research institutions, civil society and the media met in Apia for the fourth regional round table meeting on climate change (Fairbairn, 2008; Ronneberg, 2009). A key purpose of this and previous round table meetings was for countries and donors to exchange information about existing and planned projects relating to climate change in the region. At the 2008 meeting a matrix listing most ongoing and planned activities on climate change in the region was developed. This matrix updated an earlier one prepared by the World Bank in 2004, which showed that there were about 90 projects in existence or planned for the region that had some bearing on the issue of climate change. This included projects on cognate issues such as water resources management, energy planning and meteorological services, and it included very small (in value) projects, implemented by regional agencies and NGOs. The matrix that was revised in late 2008 similarly included a large number of related but not climate change specific projects (which is necessary for donor coordination), but restricted its list to include only those projects funded by multilateral and bilateral donors, thereby excluding a number of the smaller projects that were listed in the 2004 matrix. The information in the 2008 matrix is summarized in Table 6.1.

Table 6.1 *Summary of past, present and planned climate change and related projects in the South Pacific funded or implemented by major agencies*

Funding agency	Total number of projects	Adaptation	Mitigation	Technology transfer	Financial resources
Asian Development Bank	8	4	4		
AusAid	21	16	1		4
European Union	27	15	11		1
Germany	2	2			
JICA	10	8	1	1	
NZAid	15	15			
UNDP	17	7	10		
Total	100	67	27	1	5

The information in Table 6.1 shows that, by number, there are more projects on adaptation than on mitigation, technology transfer or finance. This suggests that the demands from Pacific climate change negotiators for more adaptation projects may be misplaced. However, the table should be read with caution. The number of projects does not indicate the value of the activities, or the number of countries engaged in a project (which is important given that regional projects often include many countries, yet in some cases they are counted as a single project). For example many of the projects funded by the EU have relatively small budgets, and are in single countries, whereas many of the projects funded by AusAid are regional projects implemented in multiple countries, and have larger amounts of funding associated with them. Further, the categories into which projects have been grouped (which are as they are presented in the matrix, and not our interpretation) are somewhat misleading. For example the initial national communications to the UNFCCC are categorized as adaptation projects, although they involved as much work on inventories of greenhouse gas emissions as they did on adaptation, and were principally intended to meet countries' obligations to report to the UNFCCC. Further, the matrix summarized in Table 6.1 does not include projects funded by the United Nations Environment Programme (UNEP) or the World Bank. Nevertheless, Table 6.1 does show that there are a considerable number of donor funded projects on climate change and related sustainability issues in the Pacific Islands. It also suggests that there has been an increase in climate change and related projects since the 2004 matrix was compiled.

It is instructive to examine only those projects that are about climate change *per se*, thereby excluding those projects that address related issues such as water resources management, food security, energy, and coastal and marine management. These exclusions are justified for two reasons. First, funding for adaptation should in principle be *additional* to existing development assistance, as has long been argued by the PICs and G77 in the climate change negotiations. Their argument is that if support for adaptation is not additional then support for adaptation will come at the cost of support for other activities: money for health projects, for example, may have to be redirected. The PICs argue that adaptation is a burden imposed upon them by the emissions of greenhouse gases from industrialized countries, and so there should be additional funding from these countries to help them shoulder this burden. The second reason that it is useful to exclude from the analysis projects that are not specifically about climate change is that it is hard to draw a boundary around activities that are related to adaptation, because every project can be said to contribute to adaptation in some way. For example, while the 2008 matrix includes water management projects, it could equally include all projects relating to fisheries management, sustainable livelihoods, health care and education (to name but a few), since sustainable fisheries, and a healthy and well educated population also decrease vulnerability to climate change.

Understanding the nature of climate change projects in the region entails either examining the entire set of donor funded projects in the region (a herculean task), or examining just those that are specifically about climate change. Table 6.2 presents the latter. It shows that by our reckoning there are 30 projects specifically about climate change that have been, are, or are likely to be implemented in the region. Information in this table comes from the 2004 and 2008 matrixes, our knowledge of other projects in the region, and various secondary sources. Table 6.2 excludes projects that are less than US$100,000 in value (of which we can identify approximately 12, including ten small research projects funded by Asia Pacific Network for Global Change Research (APN) [see Chapter 4]), and the WWF and Red Cross projects discussed later in this chapter). It also shows only the grant component of these projects, not the value of the associated co-financing provided by countries.

Table 6.2 shows that there has been a little over US$112 million committed by donors to climate change projects in the region. The most expensive project (the South Pacific Sea Level and Climate Monitoring project, which is discussed further in the following chapter) is also the oldest: it started in 1991, and Phase IV, which began in 2006, will last until 2010. Of all the projects shown in the table, 13 received less than US$1 million in funding. The table shows that the largest donor is the GEF, providing over US$47 million in funding. The second largest donor is Australia: AusAid has funded six projects worth US$38 million, and has contributed to two others. It is notable that most of this funding has come since the election of the Rudd government in 2007.

Classifying projects by their primary focus is an imprecise exercise, but if our categorizations are correct, then: projects investigating issues to do with climate and oceanographic science have received the largest share of funding (40 per cent); projects on adaptation have received 33 per cent of funding; projects on mitigation have received 15 per cent of funding; and projects that include both mitigation and adaptation have received 12 per cent of funding. Perhaps more accurate, and more revealing, is the classification of projects listed in Table 6.2 by their primary activity, which is shown in Table 6.3.

The data summarized in Table 6.3 support the claim by PICs that most of the funding in the region is for anything other than the actual implementation of adaptation: of the US$112 million committed to projects thus far, only US$15.6 million has been in the form of finance to support practical activities, and US$9.5 million of this is to finance sustainable energy activities. Thus, only 6 per cent of all funding to the region has been for the implementation of adaptation, giving rise to PICs' demands for support for 'concrete' adaptation activities.

Perhaps the only other notable omission from the projects discussed in this section is the GEF Small Grants Programme (GEF SGP). The GEF SGP provides small grants to communities to implement projects in any one (or more than one) of the GEF focal areas of biodiversity conservation, climate change mitigation, protection of international waters, land degradation and elimination of persistent organic pollutants. The GEF SGP operates in a number of countries in the region, but perhaps most successfully in Samoa, where approximately 30 projects have been funded thus far. There is yet to be a significant evaluation of the SGP in the Pacific, but its relatively easy application process, multiple-issues focus, and targeting of community groups makes it seem like a funding approach well worth supporting. Perhaps for these reasons both the New Zealand and Australian governments are committing funding to the GEF SGP programmes in the region.

Table 6.2 Summary of past, present and planned climate change projects in the South Pacific with a total grant value of more than US$100,000

	Project title	Major donor	Geographic focus	Budget (aprox) US$'000	Starting year	Focus	Primary activity
1	South Pacific Sea Level & Climate Monitoring Project Phases I–IV	AusAid	Region	25,000	1991	Science	Research
2	Pacific Climate Change Science Programme	AusAid	Region	15,500	2009	Science	Research
3	PACC	GEF	Region	13,000	2009	Adaptation	Capacity building, mainstreaming
4	Sustainable Energy Financing Project	GEF, Ausaid	Fiji, Marshall Islands, Papua New Guinea, Solomon Islands, Vanuatu	9500	2008	Mitigation	Financing and technical support
5	Second National Communications to UNFCCC	GEF	Region	5880	2006	Mitigation adaptation	Reporting
6	Secretariat of the Pacific Community/German Technical Cooperation (SPC/GTZ) Pacific–German Regional Programme on Adaptation to Climate Change	GTZ	Fiji, Tonga, Vanuatu	5800	2009	Adaptation	Capacity building, mainstreaming
7	Pacific Islands Greenhouse Gas Abatement through Renewable Energy Project (PIGGAREP)	GEF	Region	5225	2007	Mitigation	Technical suport, institutional strengthening, market development, finance for technology
8	Kiribati Adaptation Project Phases I and II	GEF, AusAid, NZAid	Kiribati	5100	2002	Adaptation	Capacity building, mainstreaming

Table 6.2 (*continued*)

	Project title	Major donor	Geographic focus	Budget (aprox) US$'000	Starting year	Focus	Primary activity
9	Vulnerability and Adaptation Initiative	AusAid	Region	4000	2004	Adaptation	Financing
10	Pacific Islands Climate Change Assistance Programme Phases I and II	GEF	Region	3400	1997	Mitigation adaptation	Capacity building
11	Pacific Islands Climate Predictions Project	AusAid	Region	3000	2005	Science	Capacity building
12	Pacific Future Climate Leaders Programme	AusAid	Region	3000	2009	Mitigation adaptation	Capacity building
13	Climate Change Vulnerability and Adaptation Assessment Training	GEF	Region	2700	1999	Adaptation	Capacity building
14	Capacity Support for Sustainable Management of Energy Resources in the Pacific Region	European Union	Region	1700	2008	Mitigation	Capacity building
15	Capacity Building to Develop Adaptation Measures in Pacific Island Countries (CBDAMPIC)	CIDA	Cook Islands, Fiji, Samoa, Vanuatu	1300	2002	Adaptation	Financing, capacity building
16	Regional Partnerships for Climate Change Adaptation and Disaster Preparedness	ADB	Region	1000	2007	Adaptation	Research
17	Environmental Vulnerability Index	New Zealand, Ireland, Norway	Region	1000	2000	Science	Research
18	National Adaptation Plans of Action	GEF	Kiribati, Samoa, Solomon Islands, Tuvalu, Vanuatu	970	2003	Adaptation	Reporting

Table 6.2 (continued)

	Project title	Major donor	Geographic focus	Budget (aprox) US$'000	Starting year	Focus	Primary activity
19	First National Communications to UNFCCC	GEF	Niue, Tonga, Palau	930	1998	Mitigation adaptation	Reporting
20	Promoting Climate Change Adaptation in Asia and the Pacific	Japan	Region	800	2008	Adaptation	Technical suport
21	Climate Change Adaptation Project for the Pacific	ADB	Cook Islands and Federated States of Micronesia	800	2003	Adaptation	Research
22	Samoa Community-based Adaptation Project (GEP- SGP)	GEF	Samoa	635	2008	Adaptation	Financing, capacity building
23	Regional Economic Report (Cities, Seas and Storms)	World Bank	Fiji, Kiribati	350	2000	Science	Research
24	Pacific Adaptation Programme	World Bank	Region	300	2002	Adaptation	Capacity building, mainstreaming
25	USP-NIWA Greenhouse Gas Project	USP, NIWA	Fiji	300	1994	Mitigation	Capacity building
26	USP Fiji CCA Project	AusAId	Fiji	300	2006	Adaptation	Financing, capacity building
27	Assessment of Impacts and Adaptations to Climate Change (AIACC)	GEF	Cook Islands, Fiji	220	2001	Adaptation	Research
28	Pacific Island Training Institute on Climate and Extreme Events	APN	Region	148	2004	Science	Capacity building
29	CDM feasibility studies	JICA	Region	140	2005	Mitigation	Research
30	Workshops on Climate Variability and Trends in Oceania	APN	Region	120	2000	Science	Research

Table 6.3 *Classification cf climate change projects by primary activity*

Primary activity of project	Share of all project funding
Capacity building and mainstreaming	34%
Financing	14%
Reporting	7%
Research	40%
Technical support	5%

Regional responses

It is clear from the previous section that many responses to climate change are regional-scale projects, most often delivered through regional organizations. The majority of projects listed in Table 6.2 are regional in scope, as are the majority of projects in cognate sectors such as energy and water management. Regionalism is indeed an important feature of governance in the Pacific, and it is also the case that regionalism is a defining characteristic of response to climate change in the Pacific Islands. This section discusses the role of some of the more prominent regional agencies engaged in climate change in the region, and it discusses the first major regional climate change project – PICCAP.

The CROP agencies

There are at present 11 Council of Regional Organizations in the Pacific (CROP) agencies in the Pacific (listed in Table 6.4). The first major regional organization was the South Pacific Commission (now known as the Secretariat of the Pacific Community), which was established by Australia, France, the Netherlands, New Zealand, the United Kingdom and the United States in 1947. A number of the present regional organizations emerged from the South Pacific Commission, and their prominence rose as countries in the region progressively gained their independence.

Table 6.4 *The CROP agencies*

Fiji School of Medicine (FSchM)
South Pacific Applied Geoscience Commission (SOPAC)
Pacific Islands Development Programme (PIDP)
Pacific Islands Forum Fisheries Agency (FFA)
Pacific Islands Forum Secretariat (PIFS)
Pacific Power Association (PPA)
Secretariat for the Pacific Community (SPC)
Secretariat of the Pacific Regional Environment Programme (SPREP)
South Pacific Board for Educational Assessment (SPBEA)
South Pacific Tourism Organization (SPTO)
University of the South Pacific (USP)

Regional organizations have long been a preferred conduit for metropolitan countries to engage in the region, ostensibly because they enable coordination of activities, but perhaps more honestly because they offer a single point of contact for metropolitan countries, which is easier and cheaper than engaging with 14 different countries. From the countries' point of view, one advantage of regional organizations is that they can help augment some of the problems they face from smallness. In particular, they can provide specialist skills that any given country may not be able to afford or produce endogenously. They may also arguably reduce the time and labour costs associated with having to otherwise deal with multiple countries seeking to engage on multiple issues.

Yet there is a permanent state of tension between the CROP agencies and the island countries over the degree to which the agencies serve themselves or the countries. This problem arises in part because of the heterogeneity of countries and their interests, and in part because the CROP agencies lack the core funding necessary to support countries in the ways they would like, and so are dependent on and spend much effort seeking and managing projects funded by donors, the supply of which may or may not match the demands of countries. The problem is exacerbated by the large contingent of non-Pacific Island staff in the agencies, which can be perceived as reflecting the imposition of western style values and processes on the region, and as a kind of expatriate cartel that locks out Pacific Islanders from the higher salaries that these agencies pay, the large amounts of travel that the staff in CROP agencies undertake, and the travel-related payments they receive.

There are also tensions between the donors that fund projects and the CROP agencies that implement them. Donors have demanding reporting and accounting requirements that are often not flexible enough to allow for the kinds of contingencies that arise when working in the region, their support is often determined by domestic or international political pressures rather than by country driven needs, and they are typically unwilling to allow project funding to subsidize other activities in the implementing CROP agency. Yet, the CROP agencies frequently request support from donors for projects that their member countries may not think are necessary, or as important as programme support or projects of other kinds. The CROP agencies frequently use their roles as negotiators in conventions on behalf of PICs to advocate for these projects (Turnbull, 2001).

Thus, there are some discontinuities between the demand for and supply of projects, as well as ongoing programmes of support, among donors, CROP agencies and countries. In addition to these problems, it remains that this national, regional and international activity is the business of foreigners and Pacific Island elites, and rarely do local people benefit, either through participation or the delivery of appropriate support (Watters, 1987; Turnbull, 2001). This larger pattern of relations also shapes the implementation of climate change projects in the region.

There are several regional organizations in the Pacific involved in implementing climate change responses. The Fiji School of Medicine was engaged to help prepare the Fiji component of the GEF-funded Piloting Climate Change Adaptation to Protect Human Health project, although it remains to be seen how much it will be involved in project implementation. As discussed in Chapter 5, the predominant regional body is the Pacific Islands Forum, whose role is one of high level coordination among leaders, between the region's leaders and major donors, and setting climate change issues alongside other regional concerns.

The CROP agencies most involved in responding to climate change are the University of the South Pacific (USP), the South Pacific Applied Geoscience Commission (SOPAC), the Secretariat of the Pacific Community (SPC) and the Secretariat of the Pacific Regional Environment Programme (SPREP). Of these SPREP, as the agency with the delegated responsibility for climate change activities in the region, plays the largest role, and should arguably be the only agency implementing climate change projects in the region. Coordination among these agencies – in particular among the SPC, SOPAC and SPREP – is not particularly effective, giving rise to rivalry, redundancy and inefficiency in the implementation of projects.

USP's role is perhaps the smallest, being restricted to some of the research components of larger projects (such as the Assessments of Impacts and Adaptations to Climate Change project), and training. USP continues to run the Vulnerability and Adaptation training course established under the PICCAP project, and it is implementing the Fijian Climate Change Adaptation project discussed later in this chapter. It has received several small grants from APN for climate change related research activities, and runs a number of ongoing projects related to cyclone activity and impacts, catchments and flooding, land and marine conservation, renewable energy and coastal zone management. USP was also among the first CROP agencies to warn the region about the risks of climate change. USP's most important contribution is the skilled graduates it produces, most of whom enter the workforces of the countries they come from and in so doing boost the skills base necessary for countries to adapt to climate change. USP also is the home of the Pacific Centre for Environment and Sustainable Development (PACE-SD), which was the base for the Fijian Climate Change Adaptation Project and acts as the secretariat for START-Oceania.

SOPAC, initially established to support Pacific countries in relation to possible seabed geological resources, now serves as a regional environmental scientific body. Its role has expanded to include the application of geoscience to the management and sustainable development of non-living resources (Turnbull, 2001). SOPAC runs a number of climate change and related projects and programmes, including various geographic information systems based projects, an Island Systems Management project, the Environmental Vulnerability Index (discussed in Chapter 7), the Pacific Islands Energy Policy and Strategic Action Planning (PIEPSAP) project, and projects related to water resources management, and it has a coordinating and communicating role in the South Pacific Sea Level and Climate Monitoring project (also discussed in Chapter 7). SOPAC is clearly the agency most responsible for disaster management in the region, and it has established disaster risk management plans in many countries, and continues to work with countries in disaster risk assessment and management. Given that climate change will result in increasingly intense and probably more frequent extreme events, this is a valuable contribution to adaptation in the region.

The SPC is also involved in climate change issues. The SPC sees climate change as an important cross-cutting issue affecting its responsibilities for land, marine and social resources (SPC, 2008a). The meetings of SPC's constituent countries and territories have also identified climate change as a key issue (SPC, 2007, 2008c). The SPC is also responsible for land and marine resources in the region, human resource development and public health, and these are sectors that may be adversely affected by climate change. It also supports the collection of, and houses, regional statistical (economic and demographic) data, which are important tools in impact assessment and response development. An important contribution from the SPC has been the Oceanic Fisheries and Climate Change Project (OFCCP), which has investigated the effect of climate change on tuna stocks in the region's fisheries. This has included a subproject on the socio-economic impacts of these likely changes in tuna fisheries. The SPC is now also the lead agency for a German Technical Cooperation Programme on Adaptation to Climate Change, a €4.2 million project which will mostly be implemented in Fiji.

SPREP, however, is the main agency responsible for climate change policy, projects and programmes in the region, and until recently it implemented almost all of the region's climate change projects. When it was initially established SPREP was a small unit housed in Noumea at the SPC headquarters. It became an independent intergovernmental organization in 1992 and relocated to Apia where it has grown to its present size of around 70 staff. Climate change is located in one of its two programme areas, called Pacific Futures. SPREP, like the SPC, serves all 21 of the countries and territories of the Pacific region, and the four metropolitan countries (Australia, France, New Zealand and the United States) with territories in the region are also members. SPREP's annual budget is approximately US$8 million, to which the above mentioned metropolitan countries contribute more than 95 per cent, and of which more than 90 per cent comes in the form of project funding (AusAid, 2000; Turnbull, 2001). Core funding for SPREP is minimal, and overall funding is much less than the other major CROP agencies (see Table 6.5). This creates dependency on project funding for SPREP to fulfil its commitments to countries.

Table 6.5 *CROP agency budgets*

Agency	Funding in US$ (2008)
South Pacific Applied Geoscience Commission (SOPAC)	18,980,000
Pacific Islands Forum Fisheries Agency (FFA)	14,283,262
Pacific Islands Forum Secretariat (PIFS)	23,749,429
Secretariat for the Pacific Community (SPC)	53,613,261
Secretariat of the Pacific Regional Environment Programme (SPREP)	7,763,577

SPREP's activities are broad ranging. It often acts as the conduit between funding organizations such as the GEF and PICs, and between bilateral donors and PICs. SPREP assists countries with representation at international meetings and disperses information about climate change at the regional level and manages the regional climate change round table process. In addition to this coordinating role, SPREP has and still implements a number of projects. It coordinated the PICCAP and CBDAMPIC projects (discussed below), and is coordinating and managing the PACC project. SPREP hosts the project management office for the PIGGAREP, and it coordinates the Pacific Islands Global Climate Observing System (PI-GCOS) project. It is also responsible for implementing the Pacific Island Framework for Action on Climate Change – a regional plan that covers the period 2006–2015. In 2003 SPREP prepared a resource book on climate change for use by policy makers and in education, which indicated regional, national and community based responses that may be considered to both mitigate and adapt to climate change and sea-level variability, and concluded with an overview of international responses and their implications for PICs (Hay et al, 2003). This is probably the most comprehensive overview of climate change in the region to date.

As the lead regional agency responsible for environmental matters, including climate change, SPREP is exposed to criticism. An AusAid (2000) review of SPREP identified a number of problems with respect to climate change, including that: country demands for technical assistance are unmet by supply; SPREP is preoccupied with project implementation rather than the delivery of programmes to meet the basic resource management needs of countries; SPREP implements projects that some of the larger countries feel they could implement themselves; the costs of administering projects are probably larger than the fees SPREP receives for project management, so that the small amount of core funding may well be subsidizing projects to some degree; and what SPREP does is dominated by the availability of funding from donors. In many important respects these problems arise because SPREP lacks the core funding required to meet the ongoing technical and training needs of countries, and is instead forced to chase project funding. This locks SPREP into a cycle where accountability to donors is more important than accountability to countries, a cycle that also exists in some government line agencies in the region.

Turnbull (2001, 2003), writing about SPREP's work in conservation, argues that donors support SPREP and not the countries SPREP is in turn supposed to support. She argues that SPREP and other regional agencies have normalized the involvement of external parties and their values, goals and tools in the management of environmental problems in the Pacific. This is certainly the case with respect to SPREP's climate change work, which was significantly increased with the PICCAP project discussed in the following section.

The Pacific Islands Climate Change Assistance Programme

The first major GEF-funded climate change project in the region was PICCAP. It has had a lasting legacy in the region, both for its positive outcomes, and for what people in the region understand about what donors are willing to fund, how that funding should be implemented, and the position of countries and communities in the political economy of climate change.

PICCAP's main aim was to enable PICs to fulfil their commitments under Articles 4 and 12 of the UNFCCC, which require parties to communicate, to the COP, a national inventory of anthropogenic emissions and a general description of steps taken or envisaged by the party to implement the convention. These national communications also included information about vulnerability and adaptation to climate change, as well as about technology transfer and awareness raising activities. These two articles of the convention require the COP to provide financial and technical assistance for the preparation of the communications. This funding was provided by the GEF as the major financial mechanism of the convention, and the United Nations Institute for Training and Research (UNITAR) assisted with technical support. PICCAP was the principle vehicle through which assistance was delivered to PICs to prepare the initial national communications.

The GEF allocated US$2.44 million for the first three-year phase of PICCAP, which began in July 1997. An additional US$535,000 was received from UNITAR, and a one-year extension phase was supported with an addition S$1 million from the GEF. In total, then, PICCAP received close to US$4 million in funding over four years. The project was implemented by UNDP, and executed by SPREP. It involved ten countries: the Cook Islands, the Federated States of Micronesia, Fiji, Kiribati, the Marshall Islands, Nauru, Samoa, the Solomon Islands, Tuvalu and Vanuatu. Many of PICCAP's training and regional coordination activities were open to, and attended by, personnel from Niue, Papua New Guinea and Tonga who were to prepare their national communications, but who were not formally part of PICCAP.

PICCAP was above all else a project to build capacity to report to the UNFCCC, although many of the gains extended beyond this to include increased capacity to comprehend and respond to climate change. It aimed to produce six major outputs: an inventory of greenhouse gas sources and sinks; an evaluation of mitigation options; national vulnerability assessments; an evaluation of adaptation options; a national climate change implementation plan; and the first national communication to the COP (King and Sem, 1999).

The most significant outcome of PICCAP was the establishment of climate change teams in each country – an approach recommended by UNITAR in the project design. These teams were developed by the lead national agency, and were composed of representatives from key sectors of government, civil society, other stakeholders, and in a few instances representatives from the private sector. This cross-sectoral representation was considered a useful means to help make climate change a whole of government responsibility, not just a meteorological or environmental problem. Many of these country teams were easily reformed for the preparation of the second series of national communications to the UNFCCC.

The phrase 'capacity' is used often in documents relating to PICCAP (e.g. Hay, 2000; Hay and Sem, 2000). This is unsurprising given that capacity building was the central goal of the project. But, there is at times confusion between the notion of capacity *to report to the UNFFCCC*, which was the project's goal, and capacity to do other things. The premise that underpinned PICCAP was that the capacity of countries to report to the UNFCCC was lacking. This was undoubtedly true, but the reasons why this was the case were never examined. The findings of the evaluation of PICCAP suggest some reasons why this capacity to report to the UNFCCC was lacking, and the reasons arguably still persist today (Hay, 2000). The barriers include that: meeting international obligations is less important to PICs than responding to local needs to adapt; data to support assessments was lacking; assessing greenhouse gas emissions and reporting on measures to reduce them is not a high priority for PICs given that their emissions are so small, and in contrast to their need to assess and implement adaptation; and awareness of climate change was lacking, and it was seen as a low priority by most social leaders. Moreover the skills and the kinds of formal knowledge required by the reporting process discounted the knowledge that people in the Pacific have about changes in their social and ecological systems. In short, PICs may have lacked the capacity to report to the UNFCCC because they did not prioritize the activity. Given this, that the PICs managed to submit their initial national communications indicates their commitment to the climate regime: while the reports themselves may have been of limited utility for informing national responses, they were necessary to demonstrate that the PICs take the UNFCCC seriously, have fulfilled their commitments to it, and expect other parties to do the same.

PICCAP concentrated its activities to build human capacity on the individual climate change coordinators from each country. This created problems when the project ceased and funding for the climate change coordinator positions ended. Fortunately, most coordinators remained in their countries working on climate change or related issues after the PICCAP process finished. An alternative approach might have been to direct efforts towards building a broader base of expertise in climate change to avoid dependence on individuals. This is also necessary as the climate change coordinators still have to do an impossibly wide array of tasks, ranging from climate change negotiation, project management, technical assessments of both mitigation and adaptation activities, report writing and policy development. In most OECD countries each of these tasks is conducted by a team of people, each with the respective skills required to complete each specific activity. In the Pacific the national climate change officer does most of these activities with little or no support.

PICCAP, importantly, developed the skills and confidence of the climate change officers in each country and developed country teams. However, in some instances, the capacity developed was not to complete activities the climate change officers or country teams considered to be high priorities. For example, effort was put into training in climate change negotiation, and was directed towards the compilation of greenhouse gas inventories. While both activities were necessary, these were rarely activities that the climate change officers or country teams thought were the highest priority. Thus, in his evaluation of PICCAP Hay (2000, piii) notes that:

> *There is a widespread and strong feeling in the PICCAP countries regarding the decision of the project managers to focus on the National Communication, and on participation in international negotiations related to the Convention. Such efforts have come at the expense of national activities, and particularly the implementation of substantive projects that help address climate change issues.*

Hay (2000) also notes that some countries felt the haste to submit the national communications undermined the quality of the reports and some had problems with their ownership of their contents. This haste was part of a larger problem associated with the demands imposed by UNDP to achieve project milestones, which came at the expense of quality outcomes, and which also included onerous reporting procedures that were not originally expected. He also observes that that PICCAP did not support research, nor did it build research capacity in the region. This would appear to be one of the costs of using integrated assessment models provided by outside institutions and experts, which came at the cost of funding for a full-time training officer based in SPREP, as was originally proposed.

Whether the regional approach used in PICCAP delivered better outcomes than a national communication support process organized on a country by country basis is undecided. Hay (2000, p17) argues that the regional approach was inappropriate, noting that countries did not get the level of support they expected from SPREP, and that 'the regional approach has, to some extent, resulted in a trickle-down effect – the regional needs are met first, with national activities undertaken using the remaining resources'. Future projects, Hay (2000) concludes, should focus on meeting individual country needs.

Nevertheless, it is clear that PICCAP was a success. PICCAP's immediate objective, to submit the initial national communication to the UNFCCC, was satisfied and, additionally, it built awareness of climate change in the region and enhanced the skills and confidence of a small group of people in the region, almost all of whom are still working in the region and working on climate change. PICCAP also enabled countries in the region to articulate their concerns and needs with respect to climate change and to learn about the differences between these and the concerns and needs of regional organizations, metropolitan countries, multilateral agencies, the UNFCCC and the GEF.

Projects funded bilaterally

Some of the projects listed in Table 6.2 have been funded by single countries (or groups of countries such as is the case with support from the EU) to support activities in one or more countries in the region. These are often loosely described as bilateral activities in order to distinguish them from activities funded by multilateral agencies such as the GEF or World Bank, and from activities implemented by countries themselves, or by NGOs (discussed later in this chapter). It is important to examine projects that have been funded through such bilateral assistance, as it is likely that there will be more projects of this kind in the future. For example the European Union's Global Climate Change Alliance will have Pacific components, as will Japan's Cool Earth Initiative, and the Australian government has committed A$150 million (US$120 million) until 2011 to assist the region to respond to climate change.

This section examines two adaptation-related projects that were funded by bilateral sources: the CBDAMPIC project, funded by CIDA, and the CCA project, funded by AusAid.

CBDAMPIC

The CBDAMPIC project began in 2003 and ran for three years. It was implemented by SPREP, through national climate change teams in four countries, and engaged with eight communities in the Cook Islands, three in Fiji, two in Samoa and three in Vanuatu. The project's overall aim was to develop and implement a capacity building programme to increase the ability of countries and communities to adapt to climate change. At the national level it sought to raise awareness of climate risks and to mainstream climate risk management into national planning. At the community level it sought to raise awareness of climate change risks, assess vulnerability and determine adaptation options, and implement pilot projects designed to reduce vulnerability. It also aimed to foster linkages among communities, national governments and regional agencies. The total budget for the project was US$1,390,000, although in many of the pilot projects more than half the costs of activities were met from local sources (Nakalevu, 2006).

The project heeded the lessons of PICCAP. It redefined SPREP's role as being one of providing administrative and technical support for the countries, which were primarily responsible for project implementation. This is reflected in the distribution of funding under the project, which, unlike many project documents in the region, is reported in a clear and transparent way, showing that 63 per cent of funds were spent in countries, with 18 per cent being spent on regional activities. Indeed, this devolution of responsibility went beyond the national level as many of the communities engaged in the project (in particular those in Aitutaki, Cook Islands) formed their own project teams, and the role of national agencies was to provide them with administrative and technical support. The project also worked with the national country teams established under PICCAP. In some cases the project helped to expand their membership. For example representatives from NGOs joined the Fijian country team, and a community representative joined the Samoan country team.

The project placed communities at the centre of assessments of vulnerability and adaptation. The assessment process had a substantial element of community engagement in decision making about adaptation responses, and recognized that such engagement can take many months (and up to years) in the region. It acknowledged the multiple stressors that communities face in addition to climate risks. The adaptation activities subsequently implemented in each community addressed immediate and long-term development as well as environmental needs. Many of these concerned improving access to fresh water: water tanks were installed in households in Aitutaki; water storage capacity was nearly doubled in one Fijian village and rainwater tanks and new pipes were installed in another; and freshwater springs were restored and upgraded in one of the Samoan villages. The most widely reported adaptation activity of the project was the relocation of a community of around 100 people on the small island of Tegua, in the far north of Vanuatu, away from the coast to higher ground. The Tegua community was being affected by localized coastal changes resulting in land loss, saline contamination of water sources, and water logging of surfaces throughout the village, with associated impacts on population health. The CBDAMPIC project helped the community to overcome some of the financial and administrative barriers to relocation to a site 500 metres inland from the coast. The move resulted in significantly greater access to fresh water (the project included construction of water tanks) among other benefits.

In addition to these material outcomes, CBDAMPIC also led to some institutional outcomes referred to as 'mainstreaming' adaptation. In the Cook Islands, for example, adaptation was included in the National Development Plan, and climate risks were assessed as part of a review of building codes. In Fiji and Vanuatu climate change concerns have been incorporated into the environmental impact assessment process. Vanuatu has established a permanent Climate Change Unit to be financed from the national budget, thereby breaking the dependence on project financing for climate change activities. Similar institutionalization occurred at the community level; for example Samoan villages devised and committed to practices such as protecting forests in watersheds and around freshwater springs. However, the final report is cautious with respect to mainstreaming, noting that mainstreaming should not be taken to mean that the costs of all adaptations should be borne by PICs. Further, mainstreaming is not in itself a sufficient response for reducing vulnerability.

The project also engaged in an extensive and wide range of awareness-raising activities, including: dancing, plays and song competitions; the production of printed materials such as pamphlets, posters, materials for schools, and newspaper articles; and the production of documentaries, television and radio programmes. Some of these activities, such as the making of a documentary about climate change in one of the Samoan villages, served as a tool for eliciting local perspectives. These activities also raised awareness at the national and community levels, as well as among governments (many of the workshops held in association with the project were attended by ministers and senior government officials).

The lessons from the CBDAMPIC project are important for future implementation of adaptation. The project has shown that community empowerment and ownership of local solutions are possible, and should be the basis for adaptation. It has demonstrated that assessments of vulnerability should lead to suggested adaptation actions, and should be based on engagement of local communities, assessments of current vulnerability to climatic and non-climatic stressors (as well as future risks), and assessments of adaptive capacity (Sutherland et al, 2005). In this respect, the project noted that 'analysis needs to be grounded on robust cultural, ecological and socio-economic assessments of vulnerability and coping capacity, and communicated well to the people to avoid misunderstandings and internal rifts' (Nakalevu, 2006, p48). CBDAMPIC has shown that community-based approaches can take a long time – for example the three years of the project was considered to be barely adequate – and in the future project cycles will need to adjust their expectations of rates of progress when working with communities in the region (Nakalevu, 2006). It has also shown that the costs of adaptation can be made more manageable by sharing commitments among parties, including communities who can help, if not with financial resources but with their labour and resources.

The CBDAMPIC project is notable for its distinctly Pacific tone. In his final report Nakalevu (2006) acknowledges the many people in countries and communities for their contributions to the project, and he makes mention of the significance of local institutional features that are critical to enabling adaptation, such as land tenure systems, traditional decision making structures, legislation and key government departments. Prior to CBDAMPIC, recognition of the people and institutions that are particular to places and central to any meaningful understanding of, and action on, adaptation has been absent in documentation of climate change projects in the region. In many ways the project marked a shift towards a new generation of climate change projects in PICs, characterized by a focus on national and local, rather than regional, scale activities, by increased participation from communities in decision making, and by a focus on adaptation. These characteristics are now somewhat normalized in other projects, particularly those implemented by local agencies within countries, such as the USP CCA project, and the NGO-led projects discussed below.

The USP Climate Change Adaptation (CCA) project

The AusAid-funded CCA in Rural Communities of Fiji project is little known compared to the other projects reviewed in this chapter, but in many ways it has been the most successful with respect to implementing the kinds of material and institutional responses necessary to reduce vulnerability to climate change. The project was implemented by the Institute of Applied Science and the Pacific Centre for Environment and Sustainable Development (PACE-SD), both at USP (Dumaru, 2007a). It began in July 2006 and finished in mid 2009, and the total budget was approximately US$310,000 (US$103,000 per year).

The CCA project sought to pilot an integrated community-based approach to climate change adaptation in six rural communities in Fiji. The focus of activities was on water resources or coastal management issues, consistent with Fiji's draft Climate Change Policy, which identified these two issues, as well as agriculture and health as priority areas for adaptation activities (Dumaru, 2009). Like the CBDAMPIC project, the CCA project sought to implement practical interventions, but its community engagement process was far more deliberative and extensive that those in CBDAMPIC.

The CCA project established an advisory committee composed of researchers and representatives of government agencies and NGOs (Dumaru, 2007a). It selected sites through a systematic process based on criteria such as location and environmental characteristics, the effectiveness of community leaders, the level of community interest in the project, and the capacity of the project to fund activities. For each of the six selected sites an initial workshop of between 10 and 30 people was conducted to inform people about the project, and to raise awareness about climate change (Dumaru, 2009).

At the first workshop, communities identified environmental problems of concern that were likely to be exacerbated by climate change. The project provided experts (for example from SOPAC and USP) to conduct technical assessments of the causes and possible solutions of the identified environmental problems (Dumaru, 2007b). The workshops and technical assessments were an important element of the project and in some cases revealed that local knowledge was not necessarily infallible. For example, a community was cutting mangroves on the opposite bank of a river in the belief that these were the cause of river bank erosion on the village side (Dumaru, 2007b). The technical assessments were simplified, summarized and translated into Fijian, and presented to each village both in the form of an information sheet and verbally. Equally important as the assessment, was the project team's ability to communicate the findings of that assessment to local people in a way that rendered that knowledge meaningful and acceptable (Dumaru, 2009).

A second workshop in each project village was then conducted to facilitate identification of adaptation responses and the development of a Community Adaptation Plan. The plan was implemented in each village, and included works such as the installation of solar water pumps, rainwater harvesting systems, composting toilets and the planting of mangroves, and institutional measures such as the development of a watershed management plan, bans on bush burning, training in water resource infrastructure maintenance, and enlisting the support of government to assist in slowing boat traffic to prevent river bank erosion (Dumaru, 2008). The project conducted a follow-up workshop to assess and evaluate progress towards implementation of the plan. Where necessary the plan was modified, and an ongoing monitoring process was established. In doing this, the project hoped to have set in train an ongoing process of planning, monitoring and evaluation, and adjustment that will continue after the project has formally ended.

There are a number of important lessons that have emerged from the CCA project. Firstly, the significance of working in the language has been highlighted. The project notes, for example, that there is only one Fijian word for both 'climate' and 'weather', which raises the possibility that discussions of climate change have hitherto been misconstrued as being about 'weather' by many Fijians (Dumaru, 2009). Secondly, the project has shown that community-based approaches are effective in generating widespread community support, although it has equally found that in places where a community has had a bad experience with a previous project, or where leadership is poor, the effectiveness of the community-based approach is limited. Finally, the project found that engaging women in the community-based process is constrained by existing gender roles in villages (Dumaru, 2007b).

The CCA project has been able to mobilize materially significant support from national and provincial government agencies, showing that networking between villages and governments can help communities overcome some of the technical and financial barriers to adaptation. Further, it has shown that: both local and formal technical knowledge is required, as local knowledge may not always be technically correct; technical assessments are greatly enhanced by local input; and decision making based on either, rather than both, is less likely to be effective and locally legitimate.

Finally, the project has demonstrated the need to negotiate local power dynamics, with respect to gender, and also with respect to community expectations about the distribution of funds, and about the nature of responses. In some instances villagers felt that payments to the project's community liaison officer should have been distributed across a broader range of people. In other cases people expected that adaptation would entail expensive capital (and presumably labour intensive) works. Negotiating such differences with reason and in a way that maintains community acceptance is integral to community-based approaches, and is something best done by people who understand the dynamics of village life.

The CCA project is in many ways a progression of the CBDAMPIC project: the principles and processes established by CBDAMPIC were adopted and refined by the CCA project. In the future, local communities will do most adaptation in the region, as they respond to changes in the environments upon which they depend. Most of the work to enhance adaptive capacity and begin implementation of responses, therefore, needs to be directed at the community level. Both these projects offer important lessons to guide those who are to support adaptation efforts in the future. A key lesson is that a careful approach to community engagement is required, and that such approaches cannot be rushed, need to fully respect but not necessarily unquestioningly idealize local knowledges, and are best implemented by people from within the region who understand what it means to live and work in local communities. There is a danger that as donors increasingly rush to be seen to support adaptation in the region they will overlook the need to support communities, or they will not pay heed to the valuable lessons provided by the CBDAMPIC and CCA projects.

Projects implemented by NGOs

Non-governmental organizations have been actively campaigning on the issue of climate change in the Pacific since the late 1980s. Sometimes their representations of the problem are unhelpful, as discussed in Chapter 8. Few of the major NGOs that campaign on climate change have a substantial presence in the Pacific. Greenpeace has long had an office based in Suva, although their actions are largely related to climate change mitigation and they do not engage in the implementation of adaptation projects. The International Union for Conservation of Nature (IUCN) has recently established an office in Suva and seems likely to implement adaptation projects in the future. The Red Cross has assisted many countries to manage disasters, and it has built on these to develop its PCCP. The WWF, whose efforts have been mostly aimed towards protecting biodiversity in the region, has an office in Suva, and local offices in Papua New Guinea, Solomon Islands, Fiji and Cook Islands. Its global-scale Climate Witness programme has had a significant Pacific element. In this section we discuss these last two programmes as they are the only two that have had adaptation elements, and they have had some notable successes.

WWF Climate Witness

While for the most part WWF's concerns regarding climate change relate to its likely ecological effects, it also engages with the human implications of climate change. The WWF South Pacific Programme has been part of a larger WWF programme called Climate Witness, which seeks to show how climate change is a social issue by recording and communicating the observations and effects of climate change according to people from around the world. There is a climate witness from the region: Penina Moce from the island of Kabara, in the Lau Group of eastern Fiji, who has spoken in the United Nations and other international meetings. WWF South Pacific has extended its Climate Witness project to engage with the four villages on Kabara (a raised limestone island on which all settlements are located on narrow, sandy coastal plains) using a variety of participatory methodologies, to build community resilience to climate variability and change (Areki and Fiu, 2005).

The Kabara project was based out of the WWF Suva office and was implemented by a project team and a full-time project officer from the region over 2004 and 2005 at an annual cost of US$65,000 per year. It had three aims: to empower the communities to share their stories about climate change; to help communities to better understand the causes of and risks created by climate change; and to work with communities to develop community-based strategies to encourage adaptation to climate change. Its principal mechanism was community workshops that used a range of participatory techniques to elicit information and facilitate discussion such as community mapping, the documentation of seasonal calendars, and inventories of biological resources. These were later described in a publication called the *Climate Witness Community Toolkit* (McFadzien et al, 2005), which has since been used by others to engage with communities on climate change.

Those participating in the Climate Witness project identified a number of environmental changes in Kabara that are consistent with the projected impacts of climate change. Among the changes communities identified were trees blooming a month earlier than they had in the past, coral bleaching, declining fish catches and increasing cases of fish poisoning (ciguatera), increasing frequency and intensity of storms, coastal erosion and declining rainfall – this latter observation confirmed by analysis of rainfall data conducted the Fiji Meteorological service (Areki and Fiu, 2005). The decline in rainfall has resulted in water shortages and problems with food supply.

Communities then identified actions needed to address these problems, the people or groups responsible for taking these actions, and timelines for their completion. In this way each community developed an action plan, and these were published in the form of posters with text in Fijian. Actions the communities identified and undertook themselves included coastal planting, prohibitions on destructive fishing practices (such as the use of poisons), strengthening village rules on the use of rainwater and on maintenance of rainwater tanks, and monitoring of marine resources and fisheries.

As part of the planning process areas where external assistance was required were identified, and in all four villages this was to increase water storage. The project then helped broker external assistance to increase water storage (through the installation of rainwater tanks) by approaching people from Kabara living in Suva, the provincial government and the island development council, the national government and other NGOs. The project secured funds for the tanks, and the Lau Provincial Council carried out the installation.

There was also a training element in the Kabara project which entailed training in basic science to conduct surveys of sea-grasses and corals so that the community could conduct its own monitoring of resources. The project also conducted awareness-raising activities, including presentations in schools and to community groups, the development of posters and booklets, and stories, which were widely carried in the national media.

The strength of the Climate Witness project in Kabara was its extensive use of participatory tools to elicit local knowledge and to engage communities in decision making. The project has noted – as have others discussed in this chapter – that this takes time, and that the pace communities work at is slower than that with which implementing agencies might be comfortable. Yet the Climate Witness project, like the CCA project, accepted this as part of working with communities – it takes time to learn, discuss and decide in ways that engage all the constituents within a village, and the alternative is less widespread commitment. The project has resulted in material changes in the form of rainwater tanks, and institutional changes in the form of community plans and a capacity to monitor marine resources. The toolkit has been well received, and has informed subsequent projects such as the Red Cross PCCP discussed below.

Red Cross Preparedness for Climate Change Programme

The International Federation of Red Cross and Red Crescent (IFRC) Societies has had a long-standing mission to provide protection and assistance to people affected by disasters and conflicts. It has a presence in most countries, including in most PICs. Because climate change will increase the frequency and intensity of extreme events it has become a core concern of the IFRC. In 2002 the Netherlands Red Cross established the Climate Centre to, among other things, raise awareness, and develop policy and programmes in relation to climate change and disaster preparedness (IFRC, 2003). The Climate Centre has since initiated a number of programmes to pursue these aims, one of which was the PCCP, implemented in the South Pacific.

The PCCP aimed to increase the capacity of national Red Cross societies and their partners to prepare for climate change, particularly with respect to their understanding of the issue and its consequences for their work in disaster preparedness (IFRC, 2006). There were four steps to the programme, and each national society had approximately 18 months to complete them. The first step was to conduct a workshop for IFRC staff and delegates on the risks of climate change, which involved the use of external experts. The second was to analyse climate risks and their consequences for the country, leading to a report, which included recommendations on ways to integrate climate risks into the programmes of national societies. The third step was to share the lessons of step two in a regional workshop lasting five days. The last step involved developing national and local programmes that reduce the risks posed by climate change (IFRC, 2006).

In the South Pacific, Cook Islands, Solomon Islands and Tonga completed all four steps at various times between 2005 and 2008. Kiribati completed step one, and there were pilot projects related to integration of climate change and disasters in Samoa and Tuvalu, which have evolved into sustainable activities since their inception. Funding for these activities came from various sources, notably the Canada Fund, the Japanese Red Cross, the IFRC and the Climate Centre (McNaught, 2007a). The cost of completing the four-step process in each country was no more than US$23,000, largely in the form of salary support for national societies, and the costs of participants' travel to the regional workshop. The pilot projects in Samoa and Tuvalu were funded differently, at a cost of between US$15,000 and US$23,000 per year. The project was implemented by national societies, and was supported by a regional coordinator who worked part of the time on a voluntary basis, and who also served on the advisory board for the larger project run by the Climate Centre.

The pilot projects in Samoa and Tuvalu worked to raise awareness within the national societies as well as the community at large, through workshops in schools and communities, and through radio programmes. They also worked to strengthen existing relationships between the national societies and government agencies such as the meteorology departments, disaster management offices, water authorities and health departments (IFRC, 2006). Across the region more broadly all 12 of the national Red Cross societies have been exposed to climate change issues and the importance of linking work on disaster risk reduction with adaptation. Many of these countries were represented at regional forums in 2006 and 2008 organized by the IFRC.

In Tuvalu the PCCP has been added to Red Cross's existing activities to result in a wide array of risk-reduction measures. These include planting pandanus trees in exposed areas to minimize storm damage and erosion, national radio broadcasts advising of environmental health precautions such as the need to boil drinking water in times of drought, distributing VHF radios and satellite phones to outer islands to communicate in times of emergency, and pre-positioning of emergency supplies (Tuvalu Red Cross Society, 2008). The Tuvalu project has also worked hard to support existing institutions and to avoid duplication of responsibilities among government and civil society groups, arguing that clarifying the distribution of roles and responsibilities can greatly enhance the ability of small states to adapt to climate change.

In Solomon Islands the PCCP piloted and applied a participatory assessment toolkit that combined existing Red Cross approaches with the WWF Climate Witness toolkit. The revised toolkit enabled community engagement in assessing vulnerabilities and identifying response options. Local Red Cross branches were trained in its use so that assessments could be done simultaneously across the country. Another notable feature of the Solomon Islands PCCP project is its focus on youth: a young person was selected as the climate change and disaster risk management officer, and awareness raising and capacity building activities have sought to engage young people in various ways (Webb, 2008).

The Pacific PCCP has given rise to some critical insights into adaptation in the region. It has noted that dramatic media reports that overstate the dangers of climate change are as likely to lead to despondency and disempowerment as they are to practical responses (McNaught, 2007b). It has also noted that effective awareness-raising requires translating the otherwise abstract findings of climate science into clear messages about how climate change will affect people's lives and work (McNaught, 2007b). The PCCP has highlighted that young people are a powerful resource for change. Further, it has shown that strengthening existing institutions is an important first step towards building adaptive capacity, and should be considered before new institutions are created. Finally, more than any other climate change project in the region, the PCCP has highlighted the importance of addressing gender, showing that women are often excluded from formal decision making about natural resource management, development and climate change adaptation, yet they most often bear a disproportionate share of the costs of resource degradation and disasters (Lane and McNaught, 2009). The Pacific PCCP has paved the way for developing gender analysis and participatory techniques to include women in decision making about adaptation.

Conclusions

It is possible to discern an evolution in approaches to climate change in the Pacific Islands. There was an initial phase of assessments of the kinds provided under the Association of South Pacific Environmental Institutions (ASPEI) initiative described in Chapter 4, which were subsequently subsumed by modelling approaches to impact assessment. These modelling approaches dovetailed with the second phase of regional-scale capacity-building type projects, exemplified by PICCAP. The third phase, which arguably began with CBDAMPIC and has been continued by the CCA, Climate Witness and PCCP projects, has seen an increasing focus on communities and their capacities to adapt. Across these three phases there have been four distinct shifts: in scale – from regional to local; in focus – from impacts assessments to adaptation; in nature – from processes driven by experts from outside of the region, to processes driven by people within the region; and in cost – from expensive big projects to relatively cheaper smaller ones.

This is not to say that earlier phases have, or should, pass out of existence. There is still a need to understand vulnerability to climate change, and in particular social vulnerability; and there is still a need for methodological pluralism in this endeavour, which includes the use of integrated assessment models. However, the preponderance of effort should be on community-based approaches to understanding and enhancing adaptation and adaptive capacity since the problem of climate change is understood by people in the region as impacting on human well-being. Solutions will only have traction when they are integrated with existing community concerns, values, needs and aspirations.

The very existence and success of the CBDAMPIC, CCA, Climate Witness and PCCP projects is important, as they show to others who seek to assist with adaptation in the region that community-based approaches are possible and can be effective. They also show that such approaches take time and patience, are best implemented by people within countries and communities, and that material as well as institutional changes are possible. Further, they demonstrate that capacity building can be delivered at local levels, and can be directed towards specific as well as generic tasks. They show that cheap and simple field-based assessments of physical and social vulnerability and adaptive capacity are a sufficient basis for adaptation. Finally, the projects show that communities in the Pacific are not passive victims of climate change and that they have at their disposal an array of skills, institutions and material resources that can help them to adapt to climate change, and that when appropriately engaged they will deploy these in purposeful and effective ways. This is not to say that such communities are invulnerable, but rather that their vulnerability can be greatly reduced by taking their existing capabilities seriously and by strengthening them through community-based approaches.

7

Investing in Uncertainty and Vulnerability

With the exception of some elements of the Pacific Islands Climate Change Assistance Programme (PICCAP), the projects discussed in the previous chapter were uncontroversial both with respect to their research and their contribution to assisting the Pacific Islands to respond to climate change. They presented no significant science-policy conundrums to their sponsors or to people in the region. There have, however, been two very controversial climate change related projects in the region – the South Pacific Sea Level and Climate Monitoring (SPSLCM) project, and the Environmental Vulnerability Index (EVI). In this chapter we review these projects and the controversy surrounding them, showing that the controversy has arisen largely because of their perceived political functions, which have in turn given rise to questions about whether or not the projects and methodology were suitable. The controversies also reveal some of the tensions between aid donors and the Pacific Islands, and between regional organizations and PICs. They offer insights into the highly political character of research on vulnerability in the region, and the problematic nature of efforts to measure and monitor change in the Pacific.

The chapter begins by examining the Australian-funded SPSLCM, which has been informally criticized by many as being a costly investment to produce, rather than reduce, uncertainty about climate impacts – a deployment of science that suits the Australian government's position on climate change. It then examines the New Zealand-funded EVI, which has been criticized for its outcomes and questioned over its methodology and underpinning assumptions.

The South Pacific Sea Level and Climate Monitoring project

As shown in Table 6.2, the SPSLCM is now the most expensive climate change related project in the region, with a total cost to date of A$32 million (AusAid,

2007). This is not so much because of its high annual costs, but because of its long duration – the project began in 1991 and has been funded to continue until the end of 2010. The project has been funded by AusAid throughout its duration.

The project's earlier objective was 'the provision of an accurate long term record of sea level in the South Pacific for partner countries and the international scientific community, that enables them to respond to and manage related impacts' (National Tidal Centre, 2004, p2). This was later changed to be 'to generate an accurate record of variance in long-term sea level for the South Pacific and to establish methods to make [these] data readily available and usable by Pacific Island Countries' (Hall, 2006, p1). Both objectives state the long-term nature of the project and its goal of measuring sea level, but whereas the former implies that mean sea-level change is the object of measurement, the latter describes an aim of recording 'variance in long-term sea level', implying measuring fluctuations around the mean (although inevitably of course this also implies measurement of the mean). The other subtle difference between the two objectives is one of audience: the former describes a project intended for 'partner countries and the international scientific community', whereas the latter is more clearly intended towards PICs. In practice what the project does – measure sea-level changes through a network of tide gauges – has not changed, but the rationale has. This changing justification may be seen as a response to the less than enthusiastic way the project had been received by PICs until quite recently.

The project is now in its fourth phase. Phase I was an establishment phase that lasted from 1991 to 1995. Phase I also included four training workshops focused on raising awareness and promoting education on climate and sea-level rise, including technical information regarding monitoring, data interpretation and application (Turner and Holmes, 2001). During Phase II, which lasted from July 1995 to December 2000, the SEAFRAME (Sea Level Fine Resolution Acoustic Measuring Equipment) was installed in the Cook Islands, Fiji, Kiribati, Marshall Islands, Nauru, Papua New Guinea, Solomon Islands, Tonga, Tuvalu, Vanuatu and Samoa. Phases I and II were overseen by National Tidal Facility Australia (NTFA) based at Flinders University. The NTFA also began management of Phase III, which ran from January 2001 to December 2005, but in December 2003 Flinders University withdrew its support of the project and the Australian Bureau of Meteorology established the National Tidal Centre (NTC) in its place. The first of the principal objectives discussed above was that which applied to the project up until 2003, the latter amended objective is associated with the later phases managed by the NTC.

During Phase III the NTC maintained the SPSLCM network and collected and analysed data, and released the data through monthly and annual reports. It was also during Phase III that the South Pacific Applied Geoscience Commission (SOPAC) became involved in the project. SOPAC supports the surveying aspects of the project (along with Geoscience Australia), it acts as the regional archive for the data, assists in the maintenance of the equipment, and hosts the project's

Regional Coordination and its Communications Advisor. Phase III also saw the linking of continuous global positioning systems (CGPS) to ten of the SEAFRAME sites in order to measure vertical movement of the gauges to determine absolute sea-level changes. Whereas the project always had some intention to build capacity, it was not until Phase III, when its management changed to the NTC and SOPAC became involved, that capacity building became a significant element of the project. Phase IV maintains much the same management structure as the latter half of Phase III, with the NTC and SOPAC playing lead roles. Phase IV entails the continuous collection, analysis and publication of data, as well as the installation of the remaining two GCPS units in the Solomon Islands and Marshall Islands. Data is now stored and managed by the NTC and by the University of Hawaii sea level centre. As part of Phase III a replicate regional data archive was established at SOPAC.

The SPSLCM project uses the 12 SEAFRAME stations to measure changes in sea level. The SEAFRAME stations measure water level and variables of wind speed and direction, air and sea-surface temperature, and atmospheric pressure. Sea levels are recorded every six minutes. Through this method the stations can measure water level to an accuracy of 1mm. Similar water-level measurement systems have been deployed in India, New Zealand, Saudi Arabia, the UK and the US, and those used in the US have been found to have a higher resolution and improved statistics in comparison to pressure-driven gauges (Luick, 2001).

Since the NTC took over the project the data have been communicated through monthly data reports and quarterly summary reports posted on websites (including the NTC website). Country reports have also been published for the years 2005 and 2006. The NTC also oversaw changes in relation to capacity building, which aims to increase regional and local-level participation in the project, and engage in training and technology transfer, largely through the use of training workshops. People from PICs have also been trained in differential levelling and CGPS surveys (Turner and Holmes, 2001).

The Tide in Tuvalu

Although the SEAFRAME instruments measure water level with precision, it takes in excess of 20 years of measurements to estimate sea-level trends within a 95 per cent confidence interval (Hall, 2007). Because of this temporal requirement it has thus far been difficult to use the SEAFRAME data to draw conclusions about trends in sea level from the Pacific. Until the installation of the CGPS, which began in 2001, further uncertainty existed because any movements in the land at which the tide gauges were located showed up as movements of sea level (which the SEAFRAME gauges measure as movements relative to land).

Thus, there were significant uncertainties in the data collected by the SEAFRAME gauges during the period when the SPSLCM project was being managed by the NTFA. The NTFA and AusAid were major supporters of the regional conference on Climate Change, Climate Variability and Sea level Rise held in Rarotonga in 2000. The conference was one of the major activities of PICCAP, yet it came to be dominated by the NTFA, whose work was presented in many of the plenary sessions. In these sessions the SEAFRAME data were presented as being uncertain for the reasons thus discussed, but nevertheless showing no clear trend of rising sea level in the region. In the case of Tuvalu, it was reported that sea level had fallen between 1993 and 2000.

This created considerable controversy because, as many delegates pointed out vociferously, data that are uncertain cannot be said to show a trend of any kind. The problem was not helped by some disdainful body language and behaviour among some of the Australian government delegates, and some seeming deference on the part of some NTFA staff to those delegates at critical moments. Further, other criticisms of the SPSLCM project arose at the meeting, which centred on the difficulty people in the PICs had accessing data or getting satisfactory explanations of the data from the NTFA. These problems coalesced in the minds of many delegates at the conference, including more than a few from Australian and New Zealand research agencies, that Australia's support for the SPSLCM project was not to help understand sea-level trends in the region, but to control the science in order to disprove the existence of a problem, or at best to ensure that the issue of sea-level rise in the region would remain uncertain for another decade.

The controversy continued with the publication in 2001 of an article that quoted NTFA Director Wolfgang Shearer saying 'the data does not support any sea level rise at all' (Field, 2001). A subsequent article by Baliunas and Soon (2002) (titled 'Is Tuvalu Really Sinking?') used NTFA data to question the existence of a present or future sea-level rise problem in the region, and cited in particular the case of Tuvalu which had hitherto been vocal about the risks posed by sea-level rise. The article met with some strident critique from Tuvalu, whose Assistant Secretary for Environment wrote in April 2002 that:

> ... *drawing conclusions from eight years of data is an act of scientific recklessness, bordering on gross ineptitude ... It is no coincidence to us that the Tidal Facility receives funds from Australia's aid programme. We can only speculate that the statements made by Dr Shearer have been motivated by pressure from the Howard Government ... we fear that the Australian Government may be using the tide data as another weapon to deny the impacts of climate change ... We believe that there should be a public inquiry into the representation of tidal data by the National Tidal Facility.* (Laupepa, 2002, p6)

Later in 2002, and possibly in response to Tuvalu's commentary, the Deputy Director of the NTFA told a meeting in Suva that Tuvalu's concerns about climate change were wrong, saying that 'there will be no such problem ... because sea level has been falling not rising around their country' (quoted in Nunn, 2004, p57). The long standing claims of Tuvaluans that they were witnessing increasing incursions during high tide events were dismissed by data presented by the NTFA, which they said showed that Intergovernmental Panel on Climate Change (IPCC) predictions about sea-level rise in the region were wrong (Nunn, 2004).

It was at this point that other researchers began to enter into the debate. At the Pacific Islands Forum meeting in August 2002, John Hunter, from the Antarctic Cooperative Research Centre, told the leaders that the NTFA data (showing that there was no sea-level rise in Tuvalu) was wrong. He argued that the NTF record was not long enough and covered a period during which there had been a significant El Niño event, which has the effect of lowering sea levels around Tuvalu. Hunter also said 'the way in which the NTF has released figures has been unhelpful to climate scientists but very helpful to greenhouse sceptics' (quoted in Field, 2002). His analysis, using data from a pre-existing tide gauge in Tuvalu, and adjusting for the effects of the El Niño, suggested that a cautious estimate of sea level around Tuvalu was that it had changed by between -1.1 and 2.7mm/year, but that the range of uncertainty was large (Hunter, 2002). Hunter argued then, and later (2004), that the data was consistent with IPCC estimates, and could not be used to say that sea level would not rise in the future. As of June 2006, the SEAFRAME data for Tuvalu, adjusted for land movements using the CGPS data, showed an annual rate of sea-level rise in Tuvalu of 5.7mm/year (NTC, 2006).

With the entry of Australian academics into the debate, and the carriage of that debate into the Forum, increasing pressure was placed on the NTFA, its host institution of Flinders University, and AusAid. The project was widely unpopular within Australian scientific circles and within the region. This unpopularity perhaps best explains why the project was transferred from Flinders University to the Australian Bureau of Meteorology in 2003, with the revised aims, and changes in its data sharing and capacity building elements. Since this change the SPSLCM project has been far less controversial.

So, the data shows rising sea levels in Tuvalu. Yet sceptics have and continue to use these early statements and data by the NTFA to deny that Tuvalu is or will experience sea-level rise (e.g. Eschenbach, 2004; Gray, 2006; de Freitas, 2008). They also point to local environmental changes such as groundwater depletion and the construction of causeways as being the major causes of flooding. In at least this latter sense the sceptics are right, although for the wrong reasons. There are a number of activities that have occurred in Funafuti (the main island) that have increased problems of erosion and flooding. Some of these are associated with developments that Tuvaluans have in some way decided to implement – such as the building of causeways. However, the two most careful studies of coastal change and flooding in Funafuti show that the major driver of change has been the construction of the airstrip and a large wharf by the United States military, which established a base there during the Second World War (Webb, 2006; Yamano et al, 2007). The projects changed sediment movements around the main islet of Fongafele, and they also entailed large earthworks which have left to this

Figure 7.1 *Borrow pit, Fongafale, Funafuti, Tuvalu*

day significant 'borrow pits' – vast trenches dug below the water line which fill with brackish water (such as that shown in Figure 7.1). Thus Webb (2006, p1) concludes that 'It is likely that much of the present lagoon shoreline instability experienced in Fongafale is a continuing artefact of these profound changes made by the United States military effort through the early 1940s'.

So, the major endogenous driver of coastal change in Tuvalu was not of the Tuvaluans' doing. There are two further comments to make about the comments of climate sceptics on the case of Tuvalu. First, it is entirely unremarkable to observe that some of the drivers of vulnerability to climate change are endogenously derived. This is well understood in studies of vulnerability to climate change and disasters, as the case of New Orleans, where landuse changes in the Mississippi Delta coupled with socio-economic inequality made New Orleans and its urban poor vulnerable to Hurricane Katrina (O'Brien et al, 2005). Second, none of this is to say that Tuvalu is not vulnerable to climate change; indeed, the evidence of flooding and coastal change thus far, coupled with the trend in sea-level rise, suggests that Tuvalu is very likely to experience significant impacts of climate change in the coming decades.

Since the Australian Bureau of Meteorology took over the SPSLCM project there has been much more caution expressed from the project team about its data. The director has argued there needs to be at least 19 years of data to be able to confirm the sea-level trends, and thus it cannot yet give credible indication of sea-level trends (Hall, 2007). Noting this uncertainty, Table 7.1 shows the trends detected up until June 2006, excluding the data from the Federated States of Micronesia for which the record was too short, and adjusting the data for land movements recorded by the CGPS data. It shows that sea level seems to be rising everywhere in the region, ranging from a trend of 1.7mm/yr in Fiji to 7.0mm/yr measured at the SEAFRAME station in Tonga.

Table 7.1 *Net relative sea-level rise as measured by SEAFRAME gauges to June 2006*

Location	Net relative sea level rise through to June 2006 (mm/yr).
Fiji	1.7
Kiribati	5.3
Tonga	7.0
Vanuatu	2.2
Cook Islands	2.5
Samoa	5.4
Tuvalu	5.7
Marshall Islands	4.4
Nauru	6.5
Solomon Islands	6.7
Papua New Guinea	6.2

Source: NTC, 2006

Reflecting on the SPSLCM

As we have argued in this book, science and politics cannot be separated. However, often science is not explicitly political but reflects the power relations and dominant discourse that exist within the political economy of various societies. Sometimes, however, the production of science is explicitly political because it is intentionally used for very political purposes. The story of the SPSLCM project is one of not unreasonable science being interpreted and communicated poorly, mostly to serve the interests of its funders. Tuvalu's comments about the political purposes to which the SEAFRAME data was used in 2000–2002 were apposite, and there was widespread support for this view among the PICs. The combined effect of protests from the PICs, and the response of Australian scientists such as John Hunter criticizing the NTFA's treatment of the data, resulted in a new science-policy framing for the project. The critical changes that occurred when the Bureau of Meteorology took over the project were not in the way the data was collected, but in the way it was deployed, the disposition of the staff, the sharing of roles with regional agencies and the improvements in capacity building.

The NTFA's statements about sea-level rise had purchase because of the authority that science has in climate change policy. Because the numbers they presented were 'science' the results were taken to be true despite all other evidence to the contrary. Arguments about conflicting evidence of sea-level rise based on local observations and knowledge were given little authority in Australia, and it says something about the privileged position of science that it was not until Tuvalu challenged the NTFA's analysis that the project came to be questioned. That the NTFA's analysis carried such weight says much about the authority that western science has in representing the problem of climate change in the region, and gives rise to concerns about how what is known about the region is produced and communicated. These concerns are amplified when examining the case of the EVI project.

The Environmental Vulnerability Index project

Whereas the scientific methods of the SPSLCM project were sound, its results were flawed and used quite explicitly for political purposes, and so met with resistance. In contrast, much of the concern about the EVI concerns the suitability of its methodology and underlying assumptions. Moreover, it has produced results that have significant if inadvertent political implications, and so this project too has been the subject of controversy.

Paragraph 113 of the Barbados Programme of Action for the Sustainable Development of Small Island Developing States (BPOA), states that:

> *Small island developing states, in cooperation with national, regional and international organisations and research centres, should continue work on the development of vulnerability indices and other indicators that reflect the status of small island developing states and integrate ecological fragility and economic vulnerability. (United Nations, 1994)*

In 1998 SOPAC began developing a method for quantifying environmental vulnerability, which they said was in response to this paragraph of the BPOA. The aim of the EVI is to estimate the 'vulnerability of the environment of a country to future shocks' (Kaly et al, 2004, Executive Summary).

There are many different kinds of indexes measuring numerous types of phenomena, such as those related to health (for example the body mass index), development (for example the human development index), political stability (for example the failed states index) and sustainability (for example the environmental sustainability index). Indexes simplify complex situations, and they are used to set targets, for monitoring, and for comparing entities (Villa and McLeod, 2002). On this basis they may also be used for making and changing decisions about funding. Included among the many functions SOPAC suggested for the EVI were that it could serve as a performance indicator for donor funding, and could help determine the Least Developed Country status of countries (by allowing for non-GDP based measures of development) (Kaly et al, 2001; Kaly et al, 2003).

The EVI was developed over four phases, beginning in 1998 and ending in 2004. The first phase was six months long, and involved development of the method and examination of its initial outcomes at two meetings of experts in

1999 (Kaly et al, 1999). The meetings of experts proposed criteria for determining when the EVI would be valid, including that there be no 'redundant' indicators (that is, indicators that correlate with one or more other indicators), and that the EVI be validated by independent experts (Kaly et al, 1999). Phase II was a year long, and involved further development of the EVI, development of a database and testing of results. Phase III of the EVI began in March 2000 and was the most substantial phase of the project. It sought to 'globalize the EVI, including the establishment of a global database', and to further test the model and expose it to further peer-review (Kaly et al, 2003, p1). Phase III ended with the publication of the demonstration EVI report (Kaly et al, 2003). The 'finalization' phase culminated in the 2004 Technical Report (Kaly et al, 2004).

Overall the EVI has not been an expensive project, with the total cost of all phases totalling approximately US$1 million. It was funded by the governments of New Zealand, Norway, Ireland and Italy, with support from various global and regional agencies. The EVI team has suggested that some of the index's limitations are due to insufficient funding given to the costs of collecting or buying some of the necessary data, and the demands on staff time of reporting to donors (Pratt et al, 2002).

The difficult logic of 'environmental vulnerability'

As an *environmental* vulnerability index it is interesting to examine what meaning of this term was used in the EVI. The 'environment' is defined by the EVI project as 'those biophysical systems that can be sustained without human support' (Kaly et al, 2004, 35). Herein lies one of the EVI's problems: its definition excludes environments significantly affected by human activities such as towns, agricultural landscapes and, presumably, densely populated islands. This was debated in an expert meeting early in the project, where the experts recommended that human systems be included (Kaly et al, 1999). They were not included, because SOPAC felt this 'may lead to internal conflicts in the model', and because it was assumed that 'any damage to the environment would lead to reduced human welfare' (Kaly et al, 1999, pp25–26). As we have argued elsewhere in this book, this division between society and nature that is assumed in the EVI does not really exist in the Pacific Islands; many landscapes are the product of human actions, and the 'social' and 'natural' are deeply intertwined.

In that the EVI measures changes in local environmental conditions it risks suggesting that environmental change is the product of local behaviours. Yet in many cases it is extra-territorial processes that directly and indirectly drive environmental change. The drivers of deforestation, for example, include unequal exchanges between logging companies and local people, and much of the depletion of fisheries comes from over-harvesting by foreign fishing vessels. By measuring changes within countries, the EVI therefore diverts attention from the larger political-economic processes that drive environmental change. This in turn makes the proposal that the EVI be used by donors as an indicator of a recipient country's environmental performance problematic given that the consumption of products such as fish and paper, and pollution that occurs as a result (such as greenhouse gases) in donor countries is often a significant cause of environmental change in the Pacific (Barnett et al, 2008).

The EVI defines vulnerability as 'the extent to which the environment is prone to damage and degradation', described more precisely as 'the loss of diversity, extent, quality and function of environments' (Kaly et al, 2003, p6). Sources of damage are identified as being 'natural and human events and processes, such as the weather and pollution', although in practice the EVI is most concerned with 'larger and more intense' meteorological and geological events such as cyclones and tsunamis (Kaly et al, 2003, p7).

The EVI method

The EVI is the average of 50 'smart indicators', selected on the basis of their applicability to many countries, the degree to which they can be easily collected and understood, and their ability to measure environmental change (Kaly et al, 2004). The indicators are categorized according to the three aspects in the EVI's model of vulnerability, as shown in Table 7.2. There are 32 indicators to capture the frequency and intensity of hazardous events that a country experiences, eight to indicate the resistance of the country's environment to hazards (that is its ability to cope), and ten to indicate the vulnerability that a country has acquired (that is, internal processes of change that increase vulnerability) (Kaly et al, 2004). The EVI calculates a sub-index for each of these three categories.

Some of the indicators in the hazards category imply that the environment is a risk to itself. For example, a volcano is a hazard rather than a phenomenon that makes mountains, and a flood is a risk to a country's flood-dependent riparian species rather than a process that builds flood plains. That this is possible says something about the ambiguous nature of the environment in the EVI, revealing that what it really means by 'environment' is biota and not geomorphic processes and certainly not social systems. This is also clear in the frequent interchanging of the words 'ecosystem' and 'environment' in the EVI documentation.

Across all the 50 indicators there are only a few that can account for the significant contributions that environmental and resource management institutions can make to reducing environmental problems. Those that are included, such as 'agreements', 'marine protected areas' and '(terrestrial) reserves' focus on the state, reflect western resource management ideals and ignore the more numerous and important processes associated with customary resource management institutions. Because there are so few of these indicators, and because it is not possible to assign them higher weights (see below) in calculations, the institutional processes that mitigate some of the drivers of environmental change are very much discounted in the country EVI scores.

Table 7.2 *The EVI's indicators (numbers and categories assigned by SOPAC)*

Hazards	Resistance	Damage
1. Wind	11. Land	17. Imbalance
2. Dry	12. Dispersion	21. Introductions
3. Wet	13. Isolation	22. Endangered
4. Hot	14. Relief	23. Extinctions
5. Cold	15. Lowlands	24. Vegetation
6. Sea-surface temperatures	16. Borders	26. Fragmentation
7. Volcanos	19. Migratory species	27. Degradation
8. Earthquakes	20. Endemics	45. Density
9. Tsunamis		48. Coastal
10. Slides		50. Conflicts
18. Openness		
25. Loss of vegetation		
28. Reserves		
29. Marine protected areas		
30. Farming		
31. Fertilisers		
32. Pesticides		
33. Biotechnology		
34. Productivity overfishing		
35. Fishing effort		
36. Water		
37. SO_2		
38. Waste		
39. Treatment		
40. Industry		
41. Spills		
42. Mining		
43. Sanitation		
44. Vehicles		
46. Growth		
47. Tourists		
49. Agreements		

Source: Kaly et al, 2004

Many of the eight indicators of resistance are more about a country's exposure to hazards rather than its ability to cope with them. They also seek to measure factors that are not obviously related to vulnerability. For example it is not clear how relief, land area and lowlands are positively or negatively associated with vulnerability, and while they may increase exposure to some hazards, they may be irrelevant to others.

The data for each indicator comes in diverse forms, and so they are converted into a common seven-point scale (a scale of 1–7), which then means they can be aggregated. The conversion of raw data into the scale is problematic for various reasons, as explained in Barnett et al (2008), and data for 13 of the indicators is not widely available. There is a lack of data for indicators that are critical for understanding changes in islands. For example data on sea temperature, high winds, dry periods, wet periods, heat spells and coastal vulnerability are lacking. So, while there are data for 80 per cent of the indicators for 142 countries, it may be that the 20 per cent of data that is missing pertains to critical issues in countries. It is also the case that there is data for fewer than 80 per cent of indicators for the majority of small island developing states (SIDS), which brings into question the validity of comparing SIDS and other countries.

In terms of the seven-point scale, where an indicator is seen to be of little importance a low score (towards one) is assigned, and where it is seen to be important a high score (towards seven) is assigned. In the 2003 Demonstration EVI, even though an indicator might be seen to be non-applicable a score of one was assigned nonetheless – meaning that an irrelevant issue still contributed to a positive EVI score. This later changed, so that irrelevant indicators were not scored. Yet the change seems to be incomplete: Tuvalu still has a score for landslides, and Bhutan still has a score for coastal settlements (SOPAC, 2005). In other cases irrelevant indicators are scored higher than one, for example Kiribati has had no violent conflict in the last 50 years, yet it receives a score of five for this factor (SOPAC/UNEP, 2005).

For indicators for which there is no data a score of zero is given and in calculations of average scores the denominator is adjusted down accordingly. So, no data means that a risk factor makes no contribution to the EVI score. This leads to some slippages between observations of environmental change problems and the EVI. For example, there are no data for indicators of the risks that are known to be important in Niue, such as high winds, pesticides, fisheries effort and water resources, yet these are critical issues that are missed in the EVI, and whose absence means that the EVI score for Niue understates the risks it faces (Barnett et al, 2008).

The EVI does not assign weights to indicators to reflect their relative importance to a country. Thus the only way an issue can have a relatively stronger contribution to the final score of a country is via a higher number in the seven-point scale, and this may not adequately convey the seriousness of some risks over others in some countries. So, for example, although a tsunami in 1998 killed over 2000 people in Papua New Guinea (McSaveney et al, 2000), the EVI scores them as being as important as 'land area' (both scored two), and less important than 'isolation' (scored four) or 'borders' (scored three) in its calculation of a score for that country. Therefore, in Papua New Guinea's EVI 'isolation' is twice as important as 'tsunami', a prioritization that few people from that country would accept. Weighting could be done by experts from countries as a means for them to convey their knowledge. For example experts in atolls might elect to assign a high weight to sea temperature, or people in mountainous countries might assign a higher weight to landslides.

The EVI team was advised by the expert group that the EVI should not include redundant indicators (that is, more than one indicator that reflects a similar phenomenon). Nevertheless, there are many redundant indicators in the EVI. For example, the indicator called 'coastal settlements' correlates with 15 other indicators (Kaly et al, 2003, p25). According to the EVI team, if all redundant indicators were eliminated only five to six indicators would be necessary

(Kaly et al, 2003, p25). Indeed, were these based on robust and universally available data this might lead to a far more useful EVI, as many of the most useful indicators, such as the Human Development Index, are based on a small number of indicators (Barnett et al, 2008).

Outcomes of the EVI

The EVI gives scores for 235 countries and territories, ranging from Singapore (score 428) to Suriname (score 211), from China (score 360) to Chad (score 217). In the Demonstration EVI (Kaly et al, 2003) countries were explicitly ranked by score, but later in the project they were only listed in alphabetical order. Nevertheless, the fact that each country has a score means comparisons are possible, and SOPAC offers a partial ranking in that countries are grouped into five categories of 'extremely vulnerable', 'highly vulnerable', 'vulnerable', 'at risk' and 'resilient' (Kaly et al, 2004). Table 7.3 shows a selection of these scores.

Table 7.3 *A selection of EVI scores*

Country	EVI score	Rank (of 235)	% of Indicators used in assessment
American Samoa	436	1	50
Singapore	428	2	92
Nauru	421	3	76
Japan	389	14	94
Netherlands	388	15	98
Italy	386	18	98
Korea	373	30	96
Tuvalu	367	35	78
United Kingdom	373	39	96
China	360	41	94
Germany	357	44	98
Spain	352	52	96
Marshall Islands	348	56	80
Denmark	345	59	98
Fiji	333	74	92
Samoa	328	81	78
Sweden	311	105	94
Niue	309	108	68
Solomon Islands	281	152	86
Honduras	273	165	90
Ethiopia	260	182	80
Papua New Guinea	251	193	94
Australia	238	208	96
Niger	208	228	80

Source: SOPAC, 2005

The various conceptual and methodological problems that underpin the EVI become aggregated in the final score, and lead to results that few would argue reflect reality. If we assess the results in terms of the EVI's stated aim of measuring the vulnerability of the environment, the country rankings produce some illogical results. For example, American Samoa is shown to be the most vulnerable country in the world, while neighbouring Samoa, which is in many respects very similar, is shown to be the 81st most vulnerable country. Singapore and Nauru are respectively ranked as the second and third most environmentally vulnerable countries in the world, although there are arguably no unaffected 'natural systems' in either country, and neither country is particularly badly affected by environmental perturbations. Papua New Guinea is ranked as the 193rd most environmentally vulnerable country, despite the large-scale clearing of tropical forests, significant impacts of mining, and coral bleaching, all of which would seem to be of far grater environmental significance than anything that happens in Singapore. In terms of people's vulnerability to environmental perturbations, the EVI suggests that people in Singapore are almost twice as environmentally vulnerable as people in war-torn and hunger ridden Congo, and that people in New Zealand are more environmentally vulnerable than those in Papua New Guinea (Barnett et al, 2008).

Probably because of the lack of hydrometeorological data for the region, the EVI results do not show that Pacific SIDS are particularly environmentally vulnerable. This is a highly contradictory outcome given that most people who study environment and development issues in the region, and many environmental agreements such as the UNFCCC, assert the special vulnerability of SIDS. The results also belie the EVI team's recommendation that the special vulnerability of SIDS be taken 'into account in regional and international processes, including adjustments and assistance as necessary' (Kaly at al, 2002, 38). Thus the EVI did not please many countries in the Pacific, and this was highlighted at the SOPAC Governing Council meeting in 2004 (SOPAC, 2004).

Reflecting on the EVI

The EVI project team claims that the EVI has predictive value, can help with national and regional planning by indicating places in need of assistance, can serve as a performance indicator for donor funding and can help determine the Least Developed Country status of countries by allowing for non-GDP based measures of development (Kaly et al, 2001; Kaly et al, 2003). Yet the project has such significant conceptual and methodological flaws that its results have the potential to be very misleading, and should not be used for these purposes.

The project was criticized by Pacific Island leaders at the SOPAC Governing Council meeting in 2004, and was not endorsed by the leaders of SIDS at the ten-year review of the BPOA at Mauritius in January 2005. This rejection by leaders could be read as reflecting its political inconvenience (because it does not show SIDS to be particularly vulnerable) rather than its scientific merit. However, as we show in the following chapter, the politics of vulnerability in the Pacific and SIDS more generally is very nuanced: leaders understand well the double-edge nature of vulnerability discourses and would be agnostic to the results of the EVI were they scientifically credible. What worries leaders is that the index is technically flawed and requires much more work to be robust enough to be a basis for allocating funding and measuring performance. So, while the EVI is presented as 'science', and so implicitly claims objectivity and authority, its flaws have been such that the political processes of endorsement have had little trouble marginalizing it from the policy process thus far.

Conclusions

Both the SPSLCM and the EVI projects have been in some way resisted by Pacific Island leaders and officials. The premature and inaccurate presentation of results from the SPSLCM when managed by the NTFA was met with resistance from the region, spearheaded by Tuvalu, who argued successfully that interpretation of data was bad science, even if the underlying data itself was valid. The EVI has also met with resistance, and here the arguments have been about the fundamental problems with its method as well as its results (the latter being compromised by the former). In both cases, then, the authority of science has been resisted.

The cases show clearly that science is not beyond reproach, and that leaders and officials in the Pacific are not without agency in the face of science. For those who would argue that science should be apolitical, these cases show that political processes are important antidotes to scientism, with the effect that science itself is improved. The efforts of Pacific Island leaders and officials in both cases has also served to avoid poor policy: the initial analysis from the SPSLCM suggested that action to reduce emissions of greenhouse gas was unnecessary, and the results of the EVI incorrectly prioritizes problems and places in ways that can lead to misdirected and inefficient responses.

8

Discourses of Danger

Pacific Island countries (PICs), particularly small and low-lying ones, are central figures in popular understandings and the politics of global warming. Because of climate change, countries like Tuvalu, which previously had very low international profiles, have appeared in numerous press articles, television documentaries and news accounts around the world. In many ways, representations of the islands as being vulnerable to climate change have been helpful in leveraging international support: the construction of the small islands as Davids fighting against the industrial and newly developing Goliaths has considerable popular appeal in the developed world. However, this support has resulted in little in the way of material action, either in the form of a reduction in greenhouse gas emissions or the implementation of programmes that would help Pacific Island governments and communities build effective adaptation strategies. At least with respect to adaptation responses, representations of the Pacific Islands as extremely vulnerable may have created the illusion that adaptation is pointless, and have denied the resilience, agency, capacity and potential that Pacific Island communities have and which could be useful elements of an adaptation response.

In this chapter, we examine the role that these discourses of extreme vulnerability have played. We use a variety of texts to inform our arguments, including international media reports, environmental NGO publications and campaigns, political statements (e.g. Declarations made by the Pacific Islands Forum) and national communications to the United Nations Framework Convention on Climate Change (UNFCCC). In particular we examine three interrelated discourses that link islands to climate change: island vulnerability, titanic states and environmental refugees. We begin, however, by discussing the foundational idea that insularity is a fundamental problem for PICs.

Insularity: Smallness and isolation

By definition, islands are land masses that are smaller than continents, so it is not surprising that the adjective 'small' is often attached to them. All independent island countries in the Pacific region identify themselves as small island developing states (SIDS). Smallness is most commonly defined in terms of either land area or population. Table 8.1 lists the world rankings of PICs based on these variables.

Table 8.1 *Indicators of smallness*

Land area			Population		
Country	World rank[a]	Band	Country	World Rank[b]	Band
Papua New Guinea	54	500000km²	Papua New Guinea	106	5,000,000
Solomon Islands	142	50000km²			1,000,000
New Caledonia	154		Fiji	157	
Fiji	155		Solomon Islands	166	500,000
Vanuatu	161		French Polynesia	180	
French Polynesia	173	10000km²	New Caledonia	182	
Samoa	174		Samoa	183	
Tonga	183	1000km²	Vanuatu	185	200,000
Kiribati	185		Tonga	189	
FSM	186		Kiribati	190	
CNMI	192	500km²	FSM	192	100,000
Palau	193		CNMI	197	
Niue	204		American Samoa	205	
Cook Islands	206		Marshall Islands	206	
American Samoa	207	200km²	Palau	218	20,000
Marshall Islands	208		Wallis and Futuna	219	
Wallis and Futuna	212		Nauru	221	
Tuvalu	224	50km²	Cook Islands	222	10,000
Nauru	225		Tuvalu	223	
Tokelau	227		Niue	NA	
			Tokelau	NA	
Total countries ranked	231		Total countries ranked	227	

[a] US Census Bureau: www.census.gov.cgi-bin/ipc/idbrank.pl

[b] http://en.wikipedia.org/wiki/List_of_countries_and_outlying_territories_by_area

Small countries came to the attention of administrators and theorists in the 1960s with the publication of Benedict's (1967) edited compilation on the *Problems of Smaller Territories*. The decolonization period had begun and some of the newly independent countries, such as Samoa, were small. At the multilateral level the issue of smallness was taken up by the United Nations Conference on Trade and Development (UNCTAD), which became the focal point for trade and development issues in small states within the UN system. Both the Commonwealth Secretariat and the EU have expressed regular concerns about the constraints facing small islands and their vulnerability (Commonwealth Secretariat, 1997; Haitink,

1998). Lino Briguglio is perhaps the most prolific contributor to the literature on smallness, which he often links to economic vulnerability (Briguglio, 1995, 1997; Briguglio and Kisanga, 2004; Briguglio et al, 2006).

Yet it is not clear what the significance of being small is. Some very small states such as Monaco and Singapore are very wealthy, while some large countries such as Ethiopia have very low GDP. Most of the aforementioned studies argue that small states are vulnerable to globalization and trade and currency shocks, because they have limited natural resources, small internal markets and diseconomies of scale. However, other studies have shown that the economies of islands are no more or less exposed in the global market place than others (Easterly and Kraay, 2000; Armstrong and Read, 2002). Moreover, while urbanization has exposed some Pacific Island people to poverty, most outer island communities, whose economies are only poorly represented by current economic models, still enjoy relatively secure, if relatively basic, livelihoods. From the perspective of climate change too, it is not clear what specific disadvantages arise from smallness, other than to have a greater coastline to land-area ratio, which may be significant in the event of sea-level rise. It is interesting that some least developed land-locked states are also singled out as being particularly vulnerable to the effects of global warming.

Smallness has also been linked to high proportional impacts from extreme events. As we showed in Chapter 2, extreme events such as cyclones certainly do affect the Pacific Islands. Yet it is also worth noting that because of the migratory nature of tropical cyclones, countries that have a long coastline (such as the south and east of the United States and the Queensland coast of Australia) are much more likely to experience a greater frequency of tropical cyclone landfalls. Measured against national GDP the impact of cyclones on small islands appears to be very large, but for the communities and individuals affected the losses are much the same as they are in the United States or Australia (indeed, in the case of Hurricane Katrina losses were much higher than most people in the Pacific have experienced). If smallness was inversely related to exposure to hazard, it could be expected that disaster statistics in the Pacific region would be skewed towards the smaller states and territories. Data from the International Federation of Red Cross and Red Crescent (IFRC) and ReliefWeb on hazard incidence, costs of disasters and fatalities, all point to the opposite: for example Papua New Guinea is by far the most disaster prone country in the region.

Often the term isolation is used in conjunction with smallness to describe PICs. Isolation is also difficult to define, let alone measure. In biological terms, isolation in the Pacific region tends to be measured in terms of distance from the Asian sources of most species. But such measures have little relevance to human communities other than in relation to the archaeology of the settlement of Pacific Islands.

In the contemporary Pacific, isolation operates at a range of scales. The region itself might be seen as isolated from the rest of the world, although it sits in the middle of the Pacific Rim, one of the more economically vibrant parts of the global economy. However, it is mostly bypassed by shipping and overshot by aviation routes between the east and western edges of the Pacific. Within the region some countries are more isolated than others. For example, Fiji occupies a central role within the region: it has a large number of embassies, many of which serve other island countries from Suva, it is the host of seven of the 11 Council of Regional Organizations in the Pacific (CROP) agencies, has regular shipping services from a number of lines, and Nadi airport is a regional hub. Other countries receive only a few flights a week, are dependent upon the Pacific Forum shipping line, and have limited health and education services, instead depending on Fiji and metropolitan countries for higher levels of service such as tertiary education and major surgery. In this respect isolation does constrain those elements of adaptive capacity – such as access to education and healthcare – by increasing their costs such that access is restricted, if not impossible, in remote areas.

There is also isolation within countries. For example, the Lau islands in eastern Fiji, once central in the traditional politics of the country, lie 300km from Suva. There are a total of 24 populated islands in the Lau islands (ranging from 1.5 to 61km² in area) supporting a total population of around 10,000, stretched some 450km from north to south, only four of which have airstrips. Thus an area, once a central part of a thriving military political order, is now isolated on the periphery of the most central country in the region.

The notion of isolation is linked to those of periphery and marginality. However, this infers that the 'centre' is known. But what is the centre that PICs are peripheral to? In social and economic terms, distance to New Zealand (itself an isolated place in many literatures, especially those centred on Europe), is likely to be much more significant than distance to London or New York for the communities in Samoa, Tonga, Niue, Cook Islands and Tokelau. Similarly, while Palau, Guam and the Northern Mariana Islands have strong political ties to the United States (indeed the latter two are territories), they have strong economic links with East Asia. Figure 8.1 shows the distances to various 'centres' around the Pacific Rim. The distances are large and only a handful of countries are less than 2000km from their nearest metropolitan centre.

Smallness and isolation are relational phenomena created by colonization and globalization. In so far as PICs have been incorporated into the global economy, isolation coupled with smallness presents costs. Imports and services are expensive, and returns on exports of a relatively small range of commodities are low. Many Pacific Islanders have overcome the constraints through migration, remittances and other strategies. It is worth noting though that isolation and smallness do not necessarily contribute to any greater or lesser level of vulnerability to climate change in islands than in communities that are similarly remote from

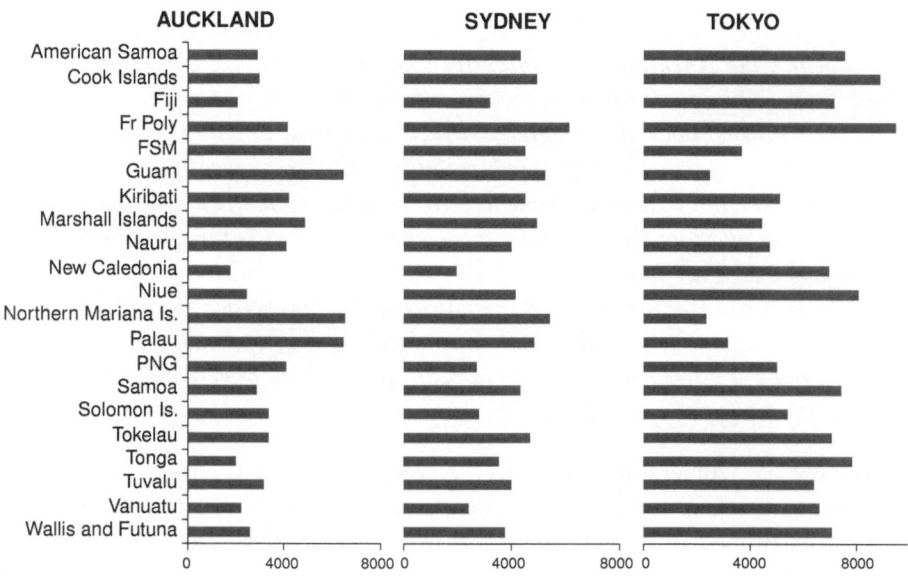

Figure 8.1 *Indicators of isolation*

economic centres on continents. In the absence of climate change there is nothing particularly problematic about smallness and isolation – most people in remote small islands live comfortable, stable and peaceful lives, but climate change may put these ways of life at risk, and distance increases the costs of building their capacity to adapt.

Hau'ofa (1993) offers a strong rebuttal of the negativism associated with economic and geographical approaches to the so-called problems of smallness and isolation in Oceania and points out that Pacific people live in a large sea of islands rather than islands in the sea. Brookfield (1980, p25) made a similar observation when he stated that 'the concept of insularity would have had as little meaning to the Pacific islanders as to the Vikings: the sea was as much of their living space as was the land.' Hau'ofa makes a compelling case that smallness is a 'belittling' state of mind. Researchers interested in island vulnerabilities could do well to take his observation into account.

What is vulnerability?

Before examining island vulnerability it is worth reflecting on the terms vulnerability and vulnerable, and the ways in which they are used. Although vulnerability has an intuitive meaning, there are many formal definitions. Vulnerability is often associated with notions such as poverty and deprivation, but it is not adequately defined by any of these terms alone. As recently as 1989, Robert Chambers, a leading writer on vulnerability, who applied the notion to those affected by famine, observed that:

> *Vulnerability, though, is not the same as poverty. It means not lack or want, but defencelessness, insecurity, and exposure to risk, shocks and stress. This contrast is clearer when different dimensions of deprivation are distinguished, for example, physical weakness, isolation, poverty, and powerlessness as well as vulnerability. Of these, physical weakness, isolation and poverty are quite well recognized, and many programmes seek to alleviate them; powerlessness is crucial but it is rare for direct action against it to be politically acceptable; and vulnerability has remained curiously neglected in analysis and policy, perhaps because of its confusion with poverty. Yet vulnerability, and its opposite, security, stand out as recurrent concerns of poor people which professional definitions of poverty overlook. (Chambers, 1989, p1)*

In its basic use, vulnerability refers to the potential for loss (Cutter, 1996). It has two key elements: an entity that is exposed, and an event or process that threatens that entity. While vulnerability arises due to some kind of external threat that can potentially cause harm to an exposed entity, it is nevertheless often the case that many of the factors that make the entity susceptible to damage lie within. This is summarized in Figure 8.2, which we call the vulnerability complex.

Watts and Bohle (1993), refine the vulnerability complex by introducing the key characteristics of capacity (to withstand or cope) and potentiality (to recover) from the manifestation of the threat. From this perspective climate change and its effects may be seen as the external element. The capacity of the people of the region to absorb or cope with the external threat and/or the potentiality to recover from it are properties that are internal to their social (including political economic)-ecological systems. If the threat does not exist, then, no matter what the conditions of the community, it cannot be vulnerable to that phenomenon. It

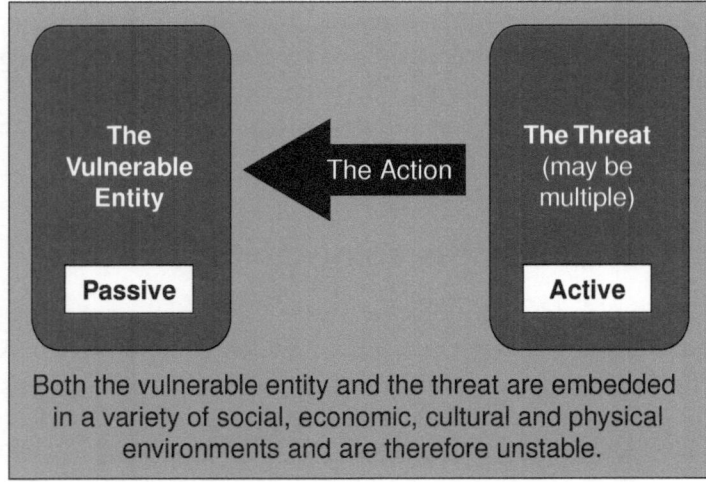

Figure 8.2 *The vulnerability complex*

follows that we cannot discuss the vulnerability of a community or a country without reference to that to which it is vulnerable. Vulnerability is therefore risk specific and is not a generic condition (a factor poorly accounted for in the EVI project discussed in the previous chapter).

We consider that there is a third element in the vulnerability complex. Usually, vulnerability is not self identified. We have been unable to find a Pacific Island language where the term vulnerability or vulnerable can be directly translated. In most cases people take the word 'weak' and translate that. Indeed, vulnerability, and vulnerable people, communities and nations are identified by experts, very few of which until recently were drawn from the communities themselves. To identify the vulnerable and to improve their situation, methods and models for vulnerability and adaptation (V&A) analyses have been developed. Reading reports by these experts (and we have been included in this work ourselves), they are typically detached (or disembodied) from the contextual environments in which the vulnerable entity and threat are located, following scientific norms. In this relationship of expert to object, the expert is presumably

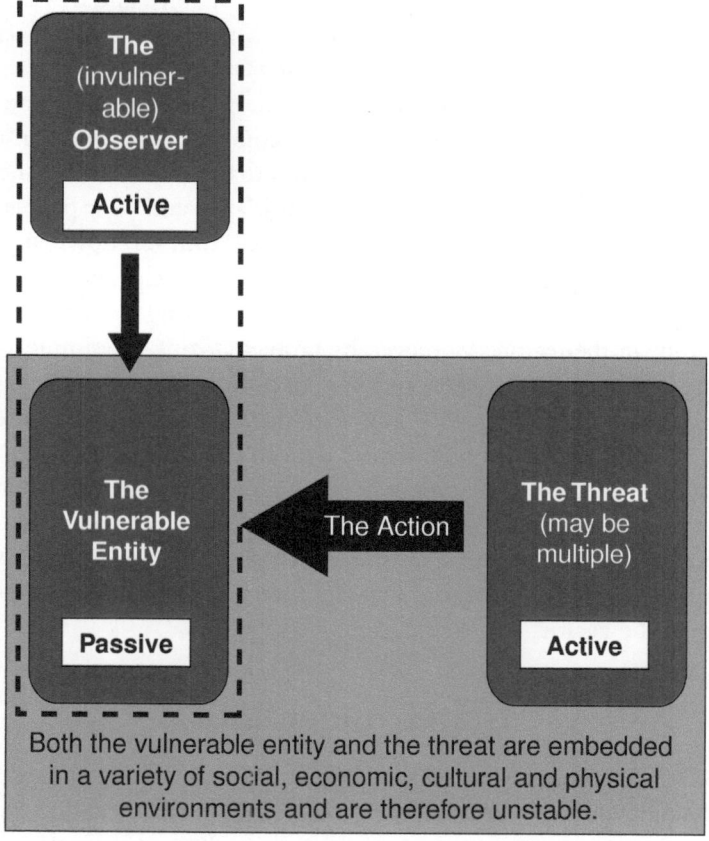

Figure 8.3 *The modified vulnerability complex*

invulnerable, while those scripted as vulnerable are generally cast as passive and incapable (see Figure 8.3). This relationship then makes it possible for the invulnerable expert to assume that he or she has the power to solve the problems of the passive and powerless group (the vulnerable). While there has been a move towards community-based analyses to help reduce this power/knowledge dynamic of vulnerability, the concept and the methods used still too frequently entail unequal relationships of power.

Among many geographers concerned with natural disasters, vulnerability to extreme events is increasingly seen as the product of social rather than natural processes. O'Keefe et al (1976) famously took the 'naturalness out of natural disaster' in an article in *Nature*. Early work by Susman et al (1983) and Watts (1983) situated susceptibility to harm from natural disasters in colonial and political-economy processes that deepened conditions of underdevelopment. More recent studies have shown how physical events are merely catalysts or triggers that bring to light communities and individuals made increasingly vulnerable through incorporation into the global economy (Watts, 1993; Wisner et al, 2004). Such studies have explanatory power, yet they do not imply that external threats are not sometimes critical elements of vulnerability, and this is particularly true in the case of climate change, which is a new threat that Pacific Island communities, among others, will need to address.

Wisner et al (2004, p19) reject 'strong social constructionist' approaches to vulnerability as they do not consider them to contribute too much in the way of solutions. The next section, however, indicates that deconstructing the taken for granted, seemingly common sense notion that islands and their people are inherently vulnerable offers opportunities to break free from top-down approaches to climate change adaptation that exude scientism, and help open up approaches that better reflect the realities of island life and the existing strengths and capabilities of people in the region. Moreover, by problematizing the simple idea that all islands are inherently vulnerable, and instead positing that they are exposed to a new threat (climate change) that is not of their doing, points to the need for measures to mitigate the problem rather than to change the behaviour of people in the Pacific. The change that most needs to come is in the behaviour of people who emit large amounts of greenhouse gases as it is this behaviour that drives the problem. This understanding stands in contrast to the now popular suggestion that vulnerability must be reduced first and foremost by changes in the behaviour of people in the Pacific.

Island vulnerability

The UNFCCC and the Kyoto Protocol identify small island states as being particularly vulnerable to climate change. Other international declarations also highlight island vulnerability: Agenda 21 and the Barbados Declaration on the Sustainable Development of Small Island Developing States identify them as being vulnerable to environmental degradation, and the Hyogo Declaration on Natural Disaster Reduction identifies them as being vulnerable to extreme events. Thus, the issue of climate change has served to strengthen and reproduce existing discourses of island vulnerability, many of which have distant origins. The difficulty is that these origins are for the most part highly environmentally deterministic in that they do not distinguish between the vulnerability of human communities and that of biological systems. As discussed in the previous chapter, it does not necessary follow that the supposed vulnerability of biological systems in islands – such as identified by MacArthur and Wilson (1967) – translates into similar levels of vulnerability in human systems. Yet this simple idea that ecological vulnerability equals social vulnerability still lies behind much of the discussion about the vulnerability of islands to climate and other changes.

Vulnerability discourses are structured around simple dualisms such as weakness/strength, risk/safety and peripherality/centrality. This is illustrated in Table 8.2. In these dualisms each side exists only in relation to the other (Plumwood, 1993). Feminists have used this framework to deconstruct discourses that sustain patriarchy, and some eco-feminists have used it to show that the treatment of nature by 'man' is little different from the treatment of women by 'man' (e.g. Seager, 1993). So, vulnerable entities are defined in terms of their opposites in the binary: things that are vulnerable are not powerful, large, robust and knowing, but are weak, powerless, and fragile and naive. These characteristics imply then that the large and powerful can and should act to help the helpless from their predicament since the vulnerable cannot by definition act to help themselves. Thus, vulnerability discourses are a form of knowledge/power: they represent the world in ways that serve the interests of power.

Table 8.2 *The dualistic framework of vulnerability*

Weak	Strong
Powerless	Powerful
Insecure / Dependent	Secure / Independent
Innocent (Naive)	Knowing
Precarious	Steady
Exposed / Unprotected	Covered / Protected
Unstable	Stable
Risk	Safety
Constrained, Limited, Restricted	Free, Unlimited, Unrestricted
Fragile	Robust
Fragmented	Coherent
Small	Large
Peripheral	Central
Marginal	Important
Female	Male (After Plumwood for example)
Uncivilized	Civilized (After Said for example)

The left hand column of Table 8.2 lists a number of synonyms of vulnerable/ vulnerability, many of which are also used to describe islands. In one of the few books devoted to the human geography of islands, Royle (2001, p39) states that, '... small islands are fragile natural systems. Their size and scale make them also problematic in physical terms for human occupation.' Moreover, the people on islands are said to face numerous constraints:

> Small islands, bounded spaces, are limited in size, in land area, in resources, in economic and population potential, in political power, by their scale. There are few if any benefits from being of small scale ... usually being small scale is simply and obviously a problem. (p42)

He goes on to observe that 'small islands are places without power' (p57), and 'powerlessness' is twice used as a subheading in the book.

These characterizations are commonly identified in representations of the Pacific Islands. Thus in writing about contemporary issues facing geographers in the Pacific region, Overton (1993, p267) notes that 'the problem of smallness is compounded by fragmentation and isolation.' He then observes:

> The three factors (smallness, fragmentation, and isolation) add up to constitute a fourth key factor: vulnerability. Limited environments are fragile environments. Without internal diversity, economic or ecological, island environments are vulnerable to a wide range of hazards, whether local (pollution, erosion, rapid population growth) or global (sea level rise, changing market prices for commodities). (p268)

Such views are not new and are often repeated. Ellen Churchill Semple (1911, p411) refers to the 'curse of restricted space'. She observes that many islands have had flourishing civilizations but few could be sustained. Vidal de la Blache (1926, p158) echoes Semple's views when he notes that while islands near to continents have been 'cradles of civilisation ... regions as restricted as these would be powerless to give to human societies the stability which would ensure them against risk of destruction.'

Islands are also represented as sites of vulnerability in popular discourse. There are many variations of cartoons showing desert islands with one or two shipwrecked souls surrounded by sharks. Similarly, many works of literature concerning islands – perhaps beginning with Robinson Crusoe – focus on themes of frailty or weakness. As J-K Gibson-Graham (1998) shows, using the work of Beer (1989), islands have commonly been represented in literature as places of natural innocence, places prone to the breakdown of social order and miniaturized extremes of social and environmental problems. These themes have been continued most recently in reality television programmes where people are placed on islands where they compete against one another to survive such that only one survivor remains. Withers (1999) argues that the Pacific Islands and their people have been represented as fragile gardens of Eden since Europeans first became aware of them:

> For Bougainville, Banks, Diderot, Cook, and Forster, Tahiti was the new Eden. For Poivre and others, Mauritius was paradise. ... The more general association between New World islands and paradisal geographies was to see them as the location of peoples as yet unfallen and as sites of natural richness. ... What was true of the Americas in the early 1600s was applicable to the Pacific in the later 1700s. Such new worlds were, at once, a paradise found, yet always on the verge of being a paradise lost through the misuse of nature and the natives there. (Withers, 1999, p84)

Such prelapsarian paradises were therefore always understood as being in a state of perpetual vulnerability, liable to fall at any point.

The conflation of islands and vulnerability persists despite numerous studies indicating that Pacific Island communities are as much characterized by resilience, innovation and survival (e.g. Bayliss-Smith et al, 1988; Campbell, 1985, 1990, 2006). Indeed, Bayliss-Smith et al (1988, p283), when reflecting on one of the most detailed research projects concerning human environment relationships on small islands (conducted in eastern Fiji) found that, 'when viewed as *societies* rather than as *ecosystems*, the small islands of eastern Fiji have shown themselves to be exceptionally resilient', concluding that as compared to continental areas, 'social and economic processes on islands are not distinctively different'.

So, while many small islands are confronted with development problems, and have experienced environmental problems for centuries, these are not necessarily because of their islandness. It not our argument that the Pacific Islands do not face serious risks arising from climate change, but we do argue that the mantra of vulnerability is problematic, can be counterproductive, and so needs to be used with caution and with a sensibility to its negative connotations.

Appropriating vulnerability

As noted in Chapter 5, one of the first Pacific Island leaders to bring attention to the vulnerability of PICs to climate change was Bikenibeu Paeniu, Prime Minister of Tuvalu, at the second World Climate Conference where he informed the assembled scientists and dignitaries that '[t]his eminent gathering ... could make the difference between Tuvalu's imminent demise and its continued existence' (Paeniu, 1991). As we observed earlier in this chapter, vulnerable groups rarely identify themselves as such. Vulnerability may be seen as a discourse like orientalism (Said, 1978) or tropicality (Arnold, 2000; Bankoff, 2001) in which an 'other' is identified in ways that illustrate the primacy of the North. However, as Paeniu's speech shows, it is not just outsiders who now represent the Pacific

Islands as vulnerable: many of the region's leaders make similar claims in international forums. Pacific Island Developing Countries in their regional report to the United Nations Conference on Environment and Development stated:

> *Many of us occupy some of the smallest habitable land areas on earth and we are vulnerable to natural and human-induced disturbance of both local and global environments. We are also bound by our extreme dependence on climate, physical characteristics and biological resources of sea and land. (SPREP, 1992, p7)*

This appropriation of vulnerability discourse by people from the region should be understood as both a reflection of what they understand of the serious risks they face and the language in which that situation is commonly described, and as a strategy to use the language of the powerful to draw attention to the need to reduce emissions and support adaptation. It is also a response to the problem of having voice in international negotiations, where there are hundreds of parties and very little time for any given one to make a point such that nuances of language and subtleties of context are sacrificed for a sound bite that is heard. Further, vulnerability is central to the financial mechanisms associated with the UNFCCC and other environmental bodies, and not using it would marginalize the SIDS from opportunities to support adaptation. So there is nothing disingenuous about the SIDS working with the language of vulnerability: they face serious risks that require international responses, and vulnerability is the language that captures those risks and facilitates engagement with power. However, vulnerability is a powerful concept that cannot always be controlled, and its use can lead to unintended consequences.

Blaming the victim

Discourses that construct islands, island communities and island countries as vulnerable to external threats – be they climate change, climatic or other natural extremes, or trade shocks – identify common causes of this vulnerability, namely their smallness, isolation and fragmentation. The emphasis is therefore very much on the internal characteristics of islands, and their inhabitants, that make them vulnerable, and it is far less on the external phenomena that threaten them.

In most attempts to define strategies to respond to climate change, therefore, the solutions have been framed as strategies for the islands to 'improve' themselves in the form of improving environmental management, improving natural disaster risk-reduction activities, improving public health services or reducing impediments to development. The trope of island vulnerability – which describes it as an internal syndrome – places the blame with the islands and their peoples. In this way attention is averted from the drivers that create the global warming that the communities have to cope with. In turn, this implies that action to reduce emissions of greenhouses gases may be delayed in favour of the development of adaptive strategies – the costs of adjustments to the externalities of the North's pollution are therefore expected to be borne first and foremost by countries like those in the South Pacific, who are expected to adjust themselves first. Moreover, if vulnerability is constructed as being largely an internal characteristic of islands, then arguably there is little moral imperative to reduce emissions (as those who are likely to suffer were vulnerable anyway).

So, vulnerability discourses run the danger of reinforcing a number of stereotypes about other people and places. After analysing newspaper reporting on Tuvalu in Australian newspapers, Farbotko (2005) highlighted the risks of island vulnerability discourses:

> ... the construction of Tuvaluans as tragic victims ... is problematic in the way it presents a particular perspective of Tuvalu, through a lens of vulnerability. ... By focusing on vulnerability, alternative discourses of adaptation for Tuvaluans are marginalized. Constructing Tuvaluan identity in terms of vulnerability can also operate to silence alternative identities that emphasize more empowering qualities of resilience and resourcefulness. (Farbotko, 2005, p289)

Vulnerability discourses also delineate island communities as weak, passive, unstable and marginal. It is a small step from here to the assumption that the invulnerable expert is the only one who can offer solutions. Uncritically used, then, vulnerability discourses may enable practices that entail powerful people *doing adaptation to* island communities, rather than enabling those countries to *do adaptation for themselves.*

Titanic states

Given the vulnerability discourse, it is not surprising that the metaphor of the Titanic has been used to describe small island states in terms of climate change (and in particular sea-level rise). Island nations, particularly those in the Pacific Ocean and the Maldives, have achieved the status of endangered species in the early 21st century imaginary, ready to sink without trace, much like the Titanic did. The notion that islands will sink beneath rising waters has been repeatedly stressed in media reporting on global warming. Moreover, many environmental organizations have used the same discursive constructions of islands and their populations to stress the urgency of obtaining a global agreement to reduced greenhouse gas emissions. Indeed, several island countries have used the metaphor in international negotiations with the same aim in mind. We have found five dominating sets of islands-at-risk discourse. These are shown in Table 8.3. All of the tropes shown in Table 8.3 focus on the problem of sea-level rise, and it is worth noting that in almost all places it will not be sea-level rise that is the primary climate change related driver of social problems in the region, rather it will be changes in the timing and magnitude of precipitation, and in the frequency and intensity of extreme events that will have the most immediate social impacts.

Table 8.3 *Discourses of islands and of climate change*

Climate change tropes for the Pacific and other Islands	Source	Date
Titanic states		
Sinking islands cling to Kyoto lifebuoy	PlanetArk	2005
	Sun-Herald,	
Sinking islands battle for climate aid cash	Sydney	2003
	Earth Island	
Sinking islands, vanishing worlds	Journal	2000
Climate refugees in a drowning Pacific	WWF-Aus	2008
Drowning islands halt effort to postpone climate change talks	Guardian, UK	2003
	The Age, Mel-	
Call for action on 'drowning' islands	bourne	2006
Keeping afloat	Schmidt	2005
Disappearing world: Global warming claims tropical island	Independent, UK	2006
High tide		
High tide in Tuvalu: In the tropical Pacific, climate change threatens to create a real-life Atlantis (Global Warming)	Price	2003
High tide: The truth about our climate crisis	Lynas	2004
Dark clouds over Paradise		
Paradise lost: Climate change forces South Sea islanders to seek sanctuary abroad	Independent, UK	2008
A sinking paradise home	Greenpeace, Aust	nd
'Paradise Lost': How climate change affects Kiribati	PBS, USA	2008
Key to the world: Kiribati, Paradise in peril	ABC, USA	2007
Coalmine canaries		
Pacific canaries	UNEP	2004
The canary is drowning: Tiny Tuvalu fights back against climate change	Corp Watch	2002
Tuvalu sinks today – The rest of us tomorrow?	Los Angeles Times	2002
Extinction and survival		
Disappearing world: Global warming claims tropical island; Rising seas ... washed an inhabitated island off the face of the earth	Independent	2006
Tuvalu Toodle – oo; '... global warming's paradisiacal ground zero.'	Outside Magazine	2002
For many, climate change kills	UPI	
Survival is not negotiable	Grist	2008

The first trope is that of sinking islands, labelled titanic states by Barnett (2005). This framework posits island countries as drowning beneath rising seas, and is summed up in the title of a Reuters' article: 'Sinking Islands Cling to Kyoto Lifebuoy' (Perry, 2005). The effect of this metaphor is to present the islands as defenceless, only able to be saved through the help of others (through the good services of the Kyoto Protocol). The metaphors of the lifebuoy and the Titanic imply the need for a lifeguard to assist the drowning swimmer, although given that it was the lifeguard who pushed the swimmer into the pool in the first place, we wonder how far the metaphor can be extended!

The high tide trope is much like that of the titanic state in that both refer to the inevitability of submersion. This trope is also used because very high tides coupled with low atmospheric pressure systems do inundate some atolls, and it is possible that rising sea levels are exacerbating this process. These sinking and tidal metaphors are linked to a grimmer one of extinction and survival. Indeed the notion of survival is not an uncommon one used in relation to human environment interrelationships in the Pacific Island region (e.g. Bayliss-Smith and Feachem, 1977). Ieremia Tabai, first president of Kiribati, used the idea of survival to describe the meaning of sustainability prior to the Earth Summit in 1992 when he stated:

> ... the idea of the exercise of sustainable development is to survive on the atolls forever ... sustainability is the idea that we can survive from day to day, and ... ever after! (SPREP, 1992, p7)

Climate change may be seen as a process that threatens the very survival of Pacific communities and even entire cultures and nations. For example, an article in *Outside* magazine refers to Tuvalu as 'global warming's paradisiacal ground zero'. In rather blasé fashion the article's headline proclaims 'Tuvalu Toodle-oo'. For the leaders of Pacific Island states, especially the atoll ones, this outcome is unacceptable, and they express disappointment at the failure of the international community to respond more urgently. President Litokwa Tomeing, President of the Marshall Islands, speaking at the United Nations General Assembly in September 2008:

> If wars have been waged to protect the rights of people to live in freedom, and to safeguard their security, why will they not be waged to protect our right to survive from the onslaught of climate change? (UN News Centre, 2008)

The idea that countries should not have to negotiate for their physical survival underpins the concerns of the atoll states.

The canary in the coalmine trope is related to the survival and extinction one. Mine canaries were expendable, and it is not surprising that this analogy has little appeal among Pacific people and is often used by them to assert their anger over failure to reduce emissions. The metaphor suggests that it will be the fate of the island countries that galvanizes international action, although it will of course be too late for the islands themselves. However, given the time lags that exist between greenhouse gas emissions, climate change and sea-level rise, it may well be that by the time the islands are seriously damaged by climate change much of the rest of the world will be unable to avoid a similar fate.

Islands are not the only icons used to galvanize support for reductions in greenhouse gas emissions; other notable ones include the polar bear (National Wildlife Federation, 2008) and the penguin (Boersma, 2008). Island communities, however, may have more capacity to adapt than these species, and it is the way these tropes ignore the possibility of adaptation that is one of their principal problems. Indeed, focusing on vulnerability and ignoring adaptation and adaptive capacity can itself exacerbate vulnerability because of the way such discourses influence perceptions of the future. If it is perceived by people in the islands that they have no possible future and may have to leave their islands, then they may cease to manage their resources in a manner that will allow for the rights of future generations (Barnett and Adger, 2003). In a similar manner, if foreign investors and aid agencies believe that islands have no future then they may cease to invest in them, which is likely to undermine capacity to adapt to climate change. Thus, Barnett and Adger (2003) argue that

> *the result of lost confidence in atoll-futures may be the end of the habitability of atolls. But this could, we argue, be brought about less by the physical impacts of climate change per se, and more by a common expectation of serious climate impacts leading to changes in domestic resource use and decreased assistance from abroad. (Barnett and Adger, 2003, p330)*

Thus, the language and metaphors that describe island states as highly vulnerable and powerless in the face of climate change may themselves be a cause of vulnerability. There is a need, therefore, for a more careful language that stresses risks, adaptation, inherent capacities and the uncertainty in assessments.

Climate refugees

In a Statement at COP6, The Hague, Netherlands, November 2000, the Hon. Teleke P. Lauti, Prime Minister, Tuvalu, said:

> *The sea is our very close neighbour. In fact, on the island where I live, Funafuti, it is possible to throw a stone from one side of the island to the other. Our islands are very low lying. When a cyclone hits us there is no place to escape. We cannot climb any mountains or move away to take refuge. It is hard to describe the effects of a cyclonic storm surge when it washes right across our islands. I would not want to wish this experience on anyone. The devastation is beyond description ... This concern is so serious for our people, that the Cabinet, in which I am a member, has been exploring the possibility of buying land in a near-by country, in case we become refugees to the impacts of climate change. (UNFCCC, 2005)*

The idea that climate change will create 'environmental refugees' has received a lot of media and scholarly attention since 2001, when prominent environmentalist Norman Myers proclaimed that climate change will cause up to 200 million more migrants by 2050. According to Myers, the vast majority of these forced migrants will come from low-lying deltas in Asia and drought prone areas of sub-Saharan Africa, but others, he states, will come from small island states (Myers, 2002).

Very few international agencies that deal with migration, including the United Nations High Commission for Refugees (UNHCR) and the International Organization for Migration (IOM), accept the use of the word 'refugee' with respect to people who may be displaced by environmental change. They argue that refugees are a particular category of migrant with a special status, as defined by the 1951 Convention Relating to the Status of Refugees, and that calling the environmentally displaced 'refugees' will compromise the existing refugee regime.

There are also problems with the discourse on environmental refugees, which like those of vulnerability mischaracterize people in the region as being unconscious actors without the power to adapt to climate change. As we have seen in Chapter 2, Pacific Island people have successfully coped with political, social, cultural, religious, economic and environmental changes for centuries, and sometimes migration is one among their many deliberate adjustment strategies. The notion of refugee also implies that responses to climate change will be ad hoc and spontaneous, whereas it is clear already that there is much planning underway, and that much can be done to ensure that responses – even if they ultimately include migration – are implemented in a careful and deliberate manner. Indeed, in the mix of measures to facilitate adaptation, labour migration may well be one constructive strategy that increases adaptive capacity (Barnett and Webber, 2009).

However, the idea that climate change or environmental change more generally is a significant driver of population movement is contested, with the refugee and migration research communities generally being somewhat ambivalent about the notion of environmentally induced population movement (Black, 2001; Castles 2002). It is difficult to distinguish between push and pull factors in a migration decision, how permanent a move may be and the degree to which a person that moved exercised choice. In the literature on climate change and migration it is assumed the climate change will force people to move by causing rapid or slow onset changes in conditions which 'force' people to move. However, separating out the purely environmental drivers of forced migration from the social processes that make people vulnerable to environmental change is difficult, and in general it is social processes that create poverty and marginality that are more important determinants of likely migration outcomes than environmental changes *per se* (Barnett and Webber, 2009). The literature also ignores the fact that much of the migration in the world is voluntary and leads to positive outcomes for migrants, the places they come from and the places they move to, suggesting in turn that in some instances migration may contribute to adaptation (Barnett and Webber, 2009).

In the Pacific Islands migration is almost entirely driven by pull factors – that is the search for access to education, employment and in some cases health care. The patterns of movement are determined largely by historical processes that have created transnational social networks, which help migrants to settle into new destinations (Morton, 2002). There is no robust evidence to suggest that people in the region are being forced to move because of climate change, even in Tuvalu, which is often erroneously reported as being a place from which climate refugees are moving (Mortreux and Barnett, 2009).

Nevertheless, the past and the present may be a poor guide to the future. As global warming increases, social ecological systems will come under increasing stress, and in some systems they may cease to be able to support existing let alone future numbers of people. In the Pacific, for example, 2°C of warming will give rise to coral bleaching every year, and the impacts of this on the fisheries that provide much of the protein that people eat may be significant, and may increase the likelihood that people will seek to move elsewhere (Donner et al, 2005). As global average temperature moves beyond a 2°C increase above pre-industrial levels, and as populations increase, the likelihood of large-scale population movements increases.

Thus we are entering a new era, where environmental degradation in many parts of the world, particularly in developing countries, and not just in islands, is increasingly likely to cause livelihoods to deteriorate, if not completely fail, and people will be increasingly impelled to move. In PICs this is most likely to be manifested in low-lying communities that will be exposed to sea-level rise, increased storm-surge intensity (exacerbated by sea-level rise), decreased access to fisheries and increased incidence of flooding. Atolls may still exist, but the Ghyben-Herzberg lens may be destroyed or polluted by the sea. However, there are barriers to movement from islands, not least of which are the cost of travelling over large stretches of water, and especially the problems of obtaining entry into new countries (in the case of international migration). With the exception of those who live in island entities that are still colonies, and those which are freely associated with metropolitan states, and one or two countries with considerable migration links to New Zealand (Samoa and Tonga), few people in the region can overcome these barriers. The more probable outcome of climate change undermining livelihoods in the Pacific may not so much be migration, but deterioration in living standards, and rising morbidity and mortality.

It has been widely reported that New Zealand is the only country that has so far committed to take environmental refugees from countries such as Tuvalu and Kiribati (*The Guardian*, 12 October 2005, p24, International News). However, this was a misinterpretation of the Pacific Access migration scheme in which small numbers of migrants (no more than 75) are able to migrate to New Zealand for work. The scheme includes Kiribati and Tuvalu among others, but its *raison d'être* is to provide some migration and work opportunities for Pacific Island people who do not have easier access to New Zealand (like those from Samoa, the Cook Islands and Tonga). It is not a response to climate change.

Community relocation is not at all without precedent in the Pacific region. Generally, however, it is most successful where the relocation is to proximate lands that the relocating community already has tenure over. The longer the distance of the move the greater are the chances of failure (Lieber, 1977; Campbell, 2008). Where proximate relocation is feasible, communities that are likely to be exposed to climate change effects might be encouraged to consider relocation as an adaptive process. If it is necessary to relocate communities, it should be proactive and planned. In most cases it will not be necessary for some time, and should only be considered as a last resort. If movements of people across international borders are ultimately necessary, lengthy and careful consultations and discussions within countries, among the countries of the region, and between PICs and metropolitan nations are necessary to minimize the psychosocial and economic impacts of such moves.

In recent years a great deal of media coverage has been given to two atolls off the coast of Bougainville in Papua New Guinea: Takuu (also known as the Mortlock Islands) and the Carteret Islands. For several decades these islands have experienced considerable erosion and numerous episodes of inundation. A number of environmental agencies and media organizations have picked up this story and used the cases to show that climate change is already forcing migration. However, there is as yet no robust evidence about the changes that may be taking place in these islands, and it seems probable that climate is not the major cause of the problems, or that it is one of several causes, which include population growth, tectonic movement, coastal structures being affected by changing processes of sediment transfer, and changing wind and wave patterns (which may or may not be due to climate change). In the absence of concrete evidence about the causes of migration from these islands, there is a danger that identifying these communities as the world's first climate refugees may give succour to climate sceptics if it is shown that climate change has little to do with the changes that seem to be underway. Nevertheless, this does not mean these places are not vulnerable to climate change.

There have been previous attempts to relocate people from Carterets to Bougainville, described by O'Collins (1990) in the following terms.

> *The problems of adapting to a new environment for which most members of the family had little or no preparation meant that the timetable for building a new Carteret Village, establishing food gardens and moving from the transit houses had to be considerably extended. Many women sat for long periods of time thinking about their island homes. On Sundays they would often risk the 20 minute walk through terrifying tall trees and bush to reach the seashore and gaze for hours out to sea towards the atolls.* (O'Collins, 1990, p259)

As the quote from O'Collins suggests, and as we have seen in Chapter 2, in the Pacific the link between communities and their customary lands and marine areas is extremely strong. If they had to move, it would be devastating for communities to leave their land, and equally devastating if communities were to be forced to give up land to migrants from elsewhere. Thus the Prime Minister of Tuvalu, Apisai Ielemia (2008: 4), said at the recent Poznan COP13 conference:

> *We are not contemplating migration ... We are a proud nation of people, we have a unique culture which cannot be relocated to somewhere else. We want to survive as a people and as a nation. And we will survive – it is our fundamental right.*

Indeed the notion of being a refugee finds little favour with PICs. McAdam and Loughry (2009), report that the term refugee sits particularly uneasily with people, civil servants and politicians in Tuvalu and Kiribati, given its connotations of helplessness, loss of dignity, fleeing from governmental persecution and being placed in camps.

So, while climate change may not bring about the volumes of involuntary migration that some predict, where climate change does force migration that people would rather have avoided, this should be considered an impact of climate change. PIC communities, and indeed numerous others in different environmental settings, have the right to inhabit their own lands, and this implies a meaningful climate change mitigation regime that achieves deep cuts in emissions, and assistance to enable PIC communities to adapt to the changes that are now unavoidable.

Conclusions

In this chapter we have considered how PICs have been cast in scientific and popular discourses as inherently vulnerable, particularly because of their smallness and isolation. According to these discourses, their vulnerability to climate change is of such a magnitude and so intractable that people will have no choice but to flee from them as refugees. We consider all of these representations as failing to recognize: the contingent nature of the notions concerned; their complexity; the range of variations in many of the parameters which reflect them within the region; and the actualities of people's lives and livelihoods on islands. As a result, solutions to the vulnerability of islands, through adaptation, are poorly conceived. Indeed, in many instances the very *possibility* of adaptation is ignored, as is the recognition that with careful assistance island communities may well be able to construct their own responses to climate change that are effective and may avoid many impacts well into the future. Thus the way islands are represented in discourses about climate change is unlikely to lead to the kinds of responses that will promote successful adaptation. Moreover, it may serve to undermine pressure to reduce emissions of greenhouse gases. An alternative language, based on notions of risk, and focusing on adaptation and adaptive capacity, may be more likely to lead to constructive outcomes.

9

Conclusions

What is known about climate change in the Pacific Islands shapes what is done to address the problem. As we have shown in this book, the empirical basis for what is known about climate change in the region is limited and piecemeal. It is produced by people from outside the region, shaped by what the institutions that fund research think is important, transfers some of the blame for vulnerability to people living in islands, and tends to ignore the risks to, and capacities of, people in the region. The gaps in this limited information are filled by speculation and hyperbole, which mischaracterizes risks, fails to recognize the agency of people in the region, ignores the significant scope for adaptation to defer or avoid climate impacts, and denies the possibility of a future for the region that is better than the present.

As such, what is being done in response to climate change is insufficient, and largely misdirected. Efforts at mitigation globally are woefully inadequate in part because the risks to places like the Pacific Islands have been described as being long-term and of an environmental nature, rather than as imminent and very serious social problems. Responses in terms of adaptation have by and large focused on producing information required by donors rather than by the people in the region who will have to make decisions about adaptation, and on building capacity to meet the expectations of donors and multilateral agencies rather than to implement adaptation in the region. The exceptions – the knowledge of, and the community-based projects implemented by, people from within the region – are significant, but by and large unrecognized and ignored.

This book has suggested that much of the science and policy on climate change in the Pacific region has worked to marginalize some of the very communities they ostensibly have been created to serve. This is an outcome of the architectures and developed-country driven approaches to science and policy approaches to climate change. The physical science of climate change has by far the lion's share of research funding, despite climate change being a process that is caused by human actions, and which will impact on – and so requires adaptive responses that are consistent with – the needs and rights and values of individuals and communities. Most adaptation work has progressed on the assumption that there is little or no need for research on the issue of adaptation, and it has ignored the human dimensions of potential impacts and adaptation and the local scales at which they are and will increasingly play out. The implication of this approach is that managerial, bureaucratic, technical and engineering responses are the core elements of adaptation, and it is these that are more likely to be funded than those that are more people centred.

We have tried in this book to present information about and analysis of what is said and done in the name of climate change in the Pacific Islands. Throughout this critical stocktaking we have sought to strike reasoned balances between evidence and speculation, science and politics, promise and peril, and the perspectives of people from both inside and outside of the region. We hope that this might contribute in some small way to a more purposeful, constructive and socially oriented approach to understanding and responding to climate change in the region, which acknowledges and seeks to supplement the knowledge and power of people in the region.

Three themes

In this section we draw together some key messages of this book by highlighting three major themes that recur throughout it. The first theme concerns the inside/outside nature of knowledge about climate change in the Pacific Islands. The second concerns the direction of policy responses, many of which thus far can be characterized as 'top-down', by which we mean their local manifestations (if any) are shaped by the expectations and resources of actors operating at larger scales – for example the projects funded by the GEF, implemented by regional agencies and facilitated by national governments. The third concerns the separation of climate change issues out from the everyday practices of people and groups in the region, and its framing as an extraordinary problem that will cause and/or requires radical departures in existing social processes – a phenomenon we loosely call 'climate exceptionalism'.

Inside and outside knowledges

As we have seen, the great majority of the global research projects that deal with climate change are poorly informed about the materialities of Pacific Island environments and the lives of their inhabitants. This is a function of the commitment of researchers to modelling approaches, which rarely include data about local conditions, let alone qualitative information about factors such as land tenure, kinship, culture and livelihoods, which are critical to understanding the risks climate change poses to people in the region, and their capacity to adapt. Such approaches privilege expert knowledge about environmental systems, and eschew local knowledge about social and ecological context. As a result assumptions are made about Pacific Islands and their communities that often homogenize and misrepresent them. These projects give rise to suggestions about adaptive actions – such as the building of sea-walls, or the relocation of communities – when there has been precious little research on what kinds of adaptation are appropriate, culturally and socially acceptable, and likely to be effective.

But this insider/outsider dynamic also operates in other, less obvious ways. For most people in the world, including people in the Pacific Islands, climate change science (and to a large extent policy) is a very arcane knowledge system. Insiders to the process have considerable power in (re)constructing this knowledge and in deciding which knowledges are included or excluded. The communities that purportedly need this information struggle to assess what is appropriate and relevant, in part because the knowledge and the language it is conveyed in is so abstract. At times this knowledge is conveyed by climate researchers, and more often by the media, in ways that generate fear and paralysis rather than constructive engagement with the problem. This power to paralyse is a function both of the stripping away of uncertainty in results and the implicit authority of science. There is therefore a need for building public understanding of climate change, its uncertainties and possibilities for response, to enable people in communities (and government) to engage meaningfully in adapting to climate change. Efforts to do this are underway – such as the Climate Change Adaptation and Climate Witness projects discussed in Chapter 6 – where local interlocutors render the science meaningful to communities.

The power of knowledge operates at a number of different levels. Many Pacific Island climate change practitioners are placed in subordinate positions when dealing with climate scientists, international civil servants, representatives of funding agencies and consultants, who have greater access to climate change knowledge than their Pacific Island counterparts. An outcome of this is that external operators can exert undue influence in decision making about projects. The subordination is sustained by the sources of funding and the greater ability of outsiders to access it, so that the production of knowledge continues to be by people from outside the Pacific Islands, producing results that decision makers from outside the islands are interested in, and eschewing local knowledges and demands for information relevant for local decisions.

Discourses on the vulnerability of the Pacific islands largely emanate from outside the region: the vulnerability of the islands is a symbol used by researchers who need problems to investigate, journalists who need problems to sell, and NGOs who need problems to solve. This is not to say that climate change does not pose serious risks to people and places in the region, nor that many of the makers of vulnerability discourses are not well intentioned. Rather, it is to say that the dominant vulnerability discourses facilitate exchanges from which people in the region rarely benefit directly. They also too often have the effect of casting people in the region as passive and unable to act until their islands sink and they are forced to flee and become refugees. This view that the Pacific Islands and their communities are in such a parlous state enables scientists and policy makers to make decisions on their behalf, which includes predetermining adaptation responses independently of any understanding of local context.

Top-down and bottom-up responses

Adaptation, to be successful, needs to operate at the scale at which most of the important decisions about social organization are made. In the Pacific, this most often means at the level of villages. Each Pacific Island community, be it mainly dependent upon subsistence agriculture and fisheries for its livelihood through to squatter settlements in the towns and cities, will have its own set of likely impacts to cope with, together with its own set of strengths and limitations to be managed in order to effectively adapt to climate-induced changes. Further, the goals of adaptation, the measures of its success and the trade-offs that may be involved can only be understood in terms of the social context in which adaptation takes place: communities value things differently and these must be taken into account if adaptation is to be effective, efficient, legitimate and equitable. It follows, then, that community-based approaches are likely to offer the most effective approach to adaptation, as these can avoid the pitfalls of externally imposed and top-down projects which underestimate local capacities and ignore local particularities.

The kinds of pledges from developed countries to assist the Pacific Islands to respond to climate change will dwarf the approximately US$100 million worth of projects the region has received thus far. If these efforts are to be efficient and effective, and meet the needs of people in the region rather than those of donors, then they must not seek to impose standard solutions implemented through regional and national agencies, but should rather provide frameworks to facilitate community-based approaches, supported by partnerships with local, national and regional institutions. Thus the lessons offered by the various projects and programmes reviewed in Chapter 6 are important. The challenge is nevertheless daunting, as there are many thousands of communities throughout the Pacific Island region in which such processes are required. Yet there remain few alternatives, and such approaches are, at least initially, inexpensive.

It will be difficult to facilitate adaptation in communities that do not see themselves at risk from climate change, who see climate change as another manifestation of the developed countries' conservation agenda, or who discount the risks because they perceive them to be distant in time in contrast to more immediate problems. There are numerous, indeed probably the great majority of, Pacific Island communities who do not consider environmental change, let alone change caused by global warming, to be a particular problem. This is not surprising, the members of most of these communities are not insiders in the dominant global climate-change knowledge system, and what information they are privy too may often seem somewhat fantastic and difficult to believe. Without contemporary evidence such as eroding shorelines or more frequent or intense cyclones and droughts, many communities are instead concerned about more immediate livelihood problems, such as the collapse of the copra industry (a cash economy mainstay for outer islands), obtaining sufficient subsistence crops to feed growing populations, the growing costs of imported goods, the costs of schooling, social tensions over land resources and intergenerational conflicts around responsibilities for family and kin. In urban areas, households worry about jobs, shelter, crime and security. It is difficult to convince the members of such communities that they should embark on activities to offset the possible future effects of climate change, and this is perhaps the most profound challenge for efforts to facilitate climate change adaptation in the region. We raise these problems of facilitating proactive adaptation because most of the projects on adaptation in the region thus far have been reactive, that is they have been in response to an environmental problem that already exists, and if this is the pattern, and all adaptation in the future is similarly reactive, then adaptation is very unlikely to be successful.

Accordingly, there is a need for new strategies that build the capacity of communities to adapt if and when responses are needed. This means approaches that result in communities being equipped to make decisions and having access to the resources and wherewithal to act when action is required. This means that awareness-raising and resource-management approaches such as those discussed in Chapter 6 are required, as are no-regrets measures to sustain and improve livelihoods, and access to education, health services, government services, clean water and sanitation, finance and information (among other things). For many PIC communities adaptation is not necessarily urgent, but what is urgent is the need to improve the foundations that will enable adaptation to happen when needed. Such measures can be well supported and enabled by global funds, multilateral agencies, regional organizations and biltateral donors (and are indeed unlikely to succeed without significant support from these organizations), but they cannot be effectively implemented by them. Implementation is a matter for communities, governments and civil society within countries.

Climate exceptionalism

It follows then that we need to situate climate change in the broader context of social, cultural, economic and environmental changes in PICs in order to make efforts to enhance adaptive capacity consistent with local needs and values. Yet the way climate change has thus far been understood, and the way projects have been implemented, makes this difficult.

Climate change has developed its own cadre of experts, from atmospheric and integrated assessment modellers through to mainstreaming adaptation planners, and all these researchers, consultants, NGOs workers, lawyers and bureaucrats seem to circulate in a science-policy bubble that at times floats far above the places where impacts will be felt and adaptations are required. The bubble is an intense and preoccupying place, where climate change tends to dominate *every*thing so that it seems like it is the paramount environmental problem, vulnerability is the most pressing social problem, adaptation should be the prevailing goal of development and resource management policies, the UNFCCC is the premier international regime, and mitigation demands blanket policy such as global carbon markets which may justify crude interventions to control local places and peoples. To those outside the bubble, climate change is simply *one* thing, to be situated alongside all other aspects of life, now and into the future. Thus, a reason why the community-based approaches discussed in Chapter 6 have been successful (and also why they have probably been somewhat ignored by climate change research and policy) is not merely because of the scale at which they work, but also because they use mainstream community development approaches that situate climate risks and responses into the larger social and ecological milieu in which communities are situated.

The standard response to overcoming climate exceptionalism in policy is to 'mainstream' adaptation into development activities. Although there are somewhat competing explanations as to what mainstreaming entails, it generally means factoring climate change risks into all policies. The experience with mainstreaming in the Pacific thus far has paralleled a similar attempt to mainstream disaster-risk reduction: in both cases the level of rhetoric, scoping reports and consultancies has considerably exceeded the level of meaningful implementation of policy change. Mainstreaming has also become a very contentious concept in the UNFCCC negotiations. Developed countries have at times suggested that developing countries should prioritize mainstreaming above other approaches to adaptation, and developing countries have interpreted this suggestion as being the imposition of a double standard (given that developed countries themselves do not perform mainstream adaptation), and as a tactic used by the wealthy countries to avoid their obligations under the UNFCCC to assist with adaptation. Ironically, however, mainstreaming has been a no less expensive exercise in the Pacific, consuming many millions of dollars (as shown in Table 6.2). Whatever the merit of the idea of mainstreaming, like so many other responses discussed in the UNFCCC, its implementation has been less than fruitful, in part because the issue has been hampered by its politicization.

Another key approach to including adaptation into existing activities is to factor climate risks into existing disaster-risk reduction activities. Awareness of the need for this is emerging in the Pacific region, but beyond the activities of the Red Cross discussed in Chapter 6, there have been no major changes in government structures responsible for disaster management. Indeed, at the Pacific regional level, adaptation and disaster-risk reduction are largely housed in separate regional organizations – SPREP and SOPAC respectively. Given this cleavage at the regional level it is not surprising that little is being achieved in individual countries (although Vanuatu has moved to house its Climate Change and National Disaster Management Offices in the one building).

Overcoming climate exceptionalism can only occur when climate change is understood as being one of a number of issues to be addressed under the general rubric of sustainable development, to be implemented by the local, national and regional institutions that exist to advance this issue. Activities to mitigate greenhouse gas emissions and to facilitate adaptation cannot be treated as special stand-alone issues requiring distinct policy frameworks and institutions. Yet the larger policy context encourages such separation: the collective set of conventions developed at the Earth Summit in 1992 address sustainable development, yet in practice the issues have become highly distinct as each has its own Convention, Secretariat and negotiating process, all of which give rise to regional and national commitments and processes that work in isolation from each other. Indeed, in small states the task of keeping abreast of each of these agreements and reporting to them consumes most of the resources of environmental and resource management agencies. Further, countries' efforts to integrate issues in the form of practical measures such as projects that address both biodiversity conservation and adaptation are hampered by competition among regional and national domestic agencies seeking to protect their income streams, and by rules associated with GEF and other funds that restrict the possibility of activities that address multiple sustainable-development issues. Thus there are institutional barriers and perverse incentives to integrating climate change responses into sustainable development processes, and these need to be overcome for effective and efficient responses to climate change in the region.

To the future

The future environments of the Pacific Islands are likely to be considerably different from those of the present. As the climate changes, the demands for adaptive action are likely to become more urgent and pressing. It is not clear yet if Pacific Island communities can adapt to these changes in order to avoid significant losses, and there are reasons to wonder if communities can continue to live in places such as atolls.

Climate changes places Pacific Island communities in an invidious position. They are dependent upon other actors to reduce emissions and impacts. Where these attempts fail (as they already have to a significant degree) they are to some extent dependent on assistance to adapt, and with this comes an increased risk of maladaptations created by poorly or hastily executed policies and programmes, responses that serve donor rather than community interests, and misunderstanding. Where adaptation fails to secure individual and community needs, rights and values, climate change will cause loss and damage, and in some cases communities may be forced to relocate or migrate.

Yet there is nothing inevitable about these outcomes. It is not too late to reduce global emissions to the degree that the climate stabilizes at a level that communities can adapt to. Achieving this will not be easy, and the Pacific SIDS will do their part in reducing emissions, and in pressuring others to do the same, using their diplomatic capabilities, political skills and moral suasion. Despite all the rhetoric about island vulnerability, which carries some moral weight in the climate negotiations, the discourse remains somewhat vague and unsubstantiated by evidence, particularly evidence about what climate change means for the communities and cultures that live in islands. Research that identifies social risks and adaptive responses has a dual value in the respect that not only does it help with adaptation policy, it can also help highlight further to people and groups that produce large amounts of greenhouse gases that their actions have dire consequences for people elsewhere in the world.

It is also not too late – indeed there is perhaps still ample time – to develop proactive responses to enable communities to adapt to climate change, so that their needs, rights and values can be indefinitely satisfied under modest rates of warming, or at worst sustained for longer under higher rates of warming. We have argued that focusing on large-scale adjustments in social systems or large-scale engineering works is likely to be a less effective and efficient approach to adaptation in the short to near term than multiple small-scale processes to build adaptive capacity at the community (both rural and urban), local government (where it exists) and national government levels. This means fostering sustainable livelihoods, empowering communities to make decisions on adapting to climate change (with assistance where necessary), and improving access to opportunities such as education, health care, finance, employment, information and decision makers.

The kinds of programmes we envisage, then, include increased funding for climate change staff working at national and provincial levels, as it is still the case that there are too few people within government with an understanding of climate change risks and responses. There is also much to be gained by increased training and funding in environmental impact assessment processes, as EIA offers the ability to screen developments for their effects on the vulnerability of people and ecosystems, and to minimize the risks climate change poses to the development itself. A system of micro-grants that remote and particularly disadvantaged communities and households can access can do much to help them improve their livelihoods and therefore their capacity to adapt.

There is also still much work to be done with respect to improving people's access to the basic services necessary to expand their capacity to adapt, including providing all communities with access to: clean energy to support the delivery of education and health care, and to reduce demands on local ecosystems for cooking and heating; basic telecommunications services, as the ability to receive and send information is an important determinant of adaptive capacity; primary healthcare, as this can dramatically minimize the risks climate change poses to health; and primary education, as education greatly enhances the capacity of individuals, households and communities to adapt. There is also a need for climate change monitoring systems that entail regular collection and analysis of basic data across a country (rather than devising elaborate systems of national-scale indicators such as the EVI). Basic data that could be collected include about:

rates of diarrhoea, ciguatera poisoning, malaria and dengue fever reported at health care stations; catches of tuna and key reef species; prices of key locally produced staples, and where markets do not exist data on harvests from selected farmers in key locations; and data on rainfall, soil moisture and water storage levels.

The Pacific region is vast, and there are many thousands of communities in need of some support for adaptation. The labour requirements seem daunting, but this is only the case if it is assumed that enabling adaptation requires skills that people in the region do not or cannot possess. The community-based projects reviewed in Chapter 6 suggest that this is not the case. It is possible and would be a very good idea to train community-level workers in sustainable livelihoods extension, including components on climate change adaptation and disaster risk reduction. These could be employed as sustainability extension officers, whose task is to regularly visit and advise remote and disadvantaged communities on sustainable development plans and practices. These officers can help build networks among communities, and between communities and governments, civil society and funding bodies. They could provide advice and assist in project design and management. They could help broker research that seeks to learn from local knowledge.

These approaches do not mean that the international community, especially the main producers of greenhouse gases, are absolved of responsibility for adaptation. The adaptation responses suggested above, and the many other possible responses, are likely to have major costs, little if any of which should be borne by communities forced to take adaptive action. Rather, it is those who have forced the need to adapt through their emissions that must bear the costs. However, this does not mean they can control the form adaptation takes. They should instead be guided by community-driven processes. Nor should they see funding as some form of aid; support for adaptation must be additional to development assistance. For their part, the PICS should insist that the delivery of support for adaptation conforms to best practices with respect to development assistance to hedge against inefficient and inappropriate responses. At times it may be that the supply of money exceeds the capacity of local institutions to use it effectively. For this reason a well governed regional trust fund for adaptation can be used to hold excess funds in order to smooth the supply of resources over time, and avoid the risks associated with projects that seek to do too much too fast.

As Chapter 2 indicated, people in the Pacific have considerable levels of knowledge about their social and ecological systems, and a demonstrated capacity to adjust to changing social and environmental circumstances. This knowledge and power, when respected and harnessed, is the key to enhancing adaptation in the region. Enhancing the power of people in the region to respond to climate change entails removing barriers and creating opportunities for action. This includes removing the financial barriers and epistemological biases that prevent the inclusion of their knowledge in climate change science. When their knowledge is put on an equal footing with the (overly) complex models and approaches developed in the existing climate change science-policy system, what is known – and so what is in turn done – about climate change in the region will improve. This includes accepting the role of traditional knowledge systems, and the knowledge of local and national administrators, and not ignoring or overriding them in science and policy processes.

The barriers to improved policy also need to be removed, and opportunities for greater autonomy need to be enabled. This includes providing access to funds for adaptation to support locally derived adaptive responses. It also includes building the capacity of decision makers in the region to evaluate, refine and select projects emanating from external organizations and institutions. This is partly a matter of training, but mostly a matter of outsiders who come to the region respecting the knowledge and sovereignty of decision makers, improving processes of consultation and engagement (which means being patient), and being willing to adjust understandings and expectations so that they align with local realities.

Pacific Island countries and the numerous communities that are situated within them have little responsibility for the environmental changes they now face. Up until a couple of hundred years ago these communities were for the most part resilient, and despite two centuries or more of subsequent radical social and economic change, they have retained considerable levels of capacity to deal with change, and maintain ways of living that are for the great majority peaceful and sustainable. Climate change, however, poses a new level of risk that not all communities may be able to manage. The challenge to the region is to adapt to sustain their needs and rights and values, and the challenge to the international community is to reduce emissions to the extent that such adaptation is able to be effective, and to support communities in the Pacific to adapt in ways that they see fit. This is possible, and anything less is unacceptable.

References

Adams, T., Dalzell, P. and Ledua, E. (1999) 'Ocean Resources', in Rapaport, M, (ed) *The Pacific Islands Environment and Society*, Bess Press, Honolulu

Adaptation Fund Board (2008) 'Draft Provisional Operational Policies and Guidelines for Parties to Access Resources from the Adaptation Fund', Third Meeting, Bonn, 15–18 September

ADB (Asian Development Bank) (2002) *The Contribution of Fisheries to the Economies of Pacific Island countries*, ADB, Manila

ADB (2004) *Key Indicators 2004: Poverty in Asia: Measurement, Estimates, and Prospects*, ADB, Manila

Adger, N., Arnell, N. and Tompkins, E. (2005) 'Successful adaptation to climate change across scales', *Global Environmental Change*, vol 15, issue 2, pp77–86

Adger, W., Paavola, J., Mace, M., and Huq, S. (eds) (2006) *Fairness in Adaptation to Climate Change*, MIT Press, Cambridge, MA

Agrawala, S. (1998) 'Context and early origins of the Intergovermental Panel on Climate Change', *Climatic Change,* vol 39, issue 4, pp605–620

Agrawala, S. (2001) 'Integration of human dimensions in climate change assessments', *2001 Open Meeting of the International Human Dimensions of Global Change Community,* Rio de Janeiro, 6–8 October

Agrawala, S. and Fankhauser, S. (eds) (2008) *Economic Aspects of Adaptation to Climate Change. Costs, Benefits and Policy Instruments,* OECD, Paris

AIACC (Assessments of Impacts and Adaptations to Climate Change) (2002) *AIACC: Annual Report 2001–2002*, AIACC, Washington DC

Allen, B. (1989) 'Frost and drought through time and space. Part 1: The climatological record', *Mountain Research and Development*, vol 9, issue 3, pp279–305

Allen, B. (1997) 'An assessment of the impact of the drought and frost in Papua New Guinea in 1997', *PLEC News and Views*, vol 9, pp21–27

Anderson, K and Bows, A (2008) 'Reframing the climate change challenge in light of post-2000 emissions trends', *Philosophical Transactions of the Royal Society A*, vol 366, pp3863–3882

Anderson, S. O., Sarma, K. M. and Sinclair, L. (2002) *Protecting the Ozone Layer: The United Nations History*, Earthscan, London

Areki, F. and Fiu, M. (2005) *Climate Witness, Report for Kabara, Lau, Fiji Islands*, World Wildlife Fund, Suva

Armstrong, H. W. and Read, R. (2002) 'The phantom of liberty?: Economic growth and the vulnerability of small states', *Journal of International Development*, vol 14, issue 4, pp435–458

Arnold, D. (2000) ' "Illusory riches": Representations of the tropical world, 1840–1950', *Singapore Journal of Tropical Geography*, vol 21, issue 1, pp6–18

ASPEI (Association of South Pacific Environmental Institutions) Task Team (1988a) *Preliminary Report, Potential Impacts of Greenhouse Gas Generated Climatic Change and Projected Sea-Level Rise on Pacific Island States of the SPREP Region*, SPC, Noumea

ASPEI Task Team (1988b) *Potential Impacts of Greenhouse Gas Generated Climatic Change and Projected Sea-Level Rise on Pacific Island States of the SPREP Region*, ASPEI, Port Moresby

AusAid (2000) *SPREP 2000: Review of the South Pacific Regional Environment Programme, Summary Report*, AusAid Quality Assurance Series no. 22, AusAid, Canberra

AusAid (2007) *Aid and The Environment – Building Resilience, Sustaining Growth: An Environment Strategy for Australian Aid*, AusAid, Canberra

Baer, P. (2006) 'Adaptation: Who Pays Whom?', in Adger, W., Paavola, J., Mace, M. and Huq, S. (eds) *Fairness in Adaptation to Climate Change*, MIT Press, Cambridge, MA

Baliunas, S. and Soon, W. (2002) 'Is Tuvalu Really Sinking?' available at www.tuvaluislands.com/news/archived/2002/2002-02-01.htm, accessed 19 June 2009

Bankoff, G. (2001) 'Rendering the world unsafe: "Vulnerability" as Western discourse', *Disasters*, vol 25, issue 1, pp19–35

Banks, G. (2002) 'Mining and the environment in Melanesia', *The Contemporary Pacific*, vol 14, issue 1, pp39–67

Barclay, K. and Cartwright, I. (2007) 'Governance of tuna industries: The key to economic viability and sustainability in the western and central Pacific Ocean', *Marine Policy*, vol 31, issue 3, pp348–358

Barker, J. (2000) 'Hurricanes and socio-economic development on Niue Island', *Asia-Pacific Viewpoint*, vol 41, issue 2, pp191–205

Barnett, J. (2001) 'Adapting to climate change in Pacific Island countries: The problem of uncertainty', *World Development*, vol 29, issue 6, pp977–993

Barnett, J. (2005) 'Titanic states? Impacts and responses to climate change in the Pacific islands', *Journal of International Affairs*, vol 59, issue 1, pp203-219

Barnett, J. (2008a) 'The effect of aid on capacity to adapt to climate change: Insights from Niue', *Political Science*, vol 60, issue 1, pp31–47

Barnett, J. (2008b) 'The worst of friends: OPEC and G77 in the climate regime', *Global Environmental Politics*, vol 8, issue 4, pp1–8

Barnett, J. (2010) 'Climate Change Science and Policy in the South Pacific, as if People Mattered', in O'Brien, K. and St. Clair, A. (eds) *Shifting the Discourse: Climate Change as an Issue of Human Security*, Cambridge University Press, Cambridge

Barnett, T. (1990) *The Barnett Report*: A summary of the Commission of Inquiry into Aspects of the Timber Industry in Papua New Guinea, The Asia Pacific Action Group, Hobart

Barnett, J. and Adger, N. (2003) 'Climate dangers and atoll countries', *Climatic Change*, vol 61, issue 3, pp321–337

Barnett, J. and Dessai, S. (2002) 'Articles 4.8/4.9 of the UNFCCC: Adverse effects and the impacts of response measures impasse', *Climate Policy*, vol 2, issue 3, pp231–239

Barnett, J. and Webber, M. (2009) Accommodating Migration to Promote Adaptation to Climate Change: A policy brief prepared for the Secretariat of the Swedish Commission on Climate Change and Development and the World Bank World Development Report 2010 team, World Bank, Washington DC, and SCCCD, Stockholm

Barnett, J., Dessai, S. and Webber, M. (2004) 'Will OPEC lose from the Kyoto Protocol?', *Energy Policy*, vol 32, issue 18, pp2077–2088

Barnett, J., Lambert, S. and Fry, I. (2008) 'The hazards of indicators: Insights from the Environmental Vulnerability Index', *Annals of the Association of American Geographers*, vol 98, issue 1, pp102–119

Barrau, J. (1965) ' "L'humide et le sec", an essay on ethnobotanical adaptation to contrastive environments in the Indo-Pacific area', *Journal of the Polynesian Society*, vol 74, issue 2, pp329–346

Baumert, K. and Kete, N. (2001) *United States, Developing Countries, and Climate Protection: Leadership or Stalemate?*, World Resources Institute, Washington DC

Bayliss-Smith, T. P. and Feachem, R. (eds) (1977) *Subsistence and Survival. Rural Ecology in the Pacific*, Academic Press, London

Bayliss-Smith, T. P., Bedford, R., Brookfield, H. C. and Latham, M. (1988) *Islands, Islanders and the World. The Colonial and Post-Colonial Experience of Fiji*, Cambridge University Press, Cambridge

Becken, S. (2005) 'Harmonising climate change adaptation and mitigation: The case of tourist resorts in Fiji', *Global Environmental Change*, vol 15, pp381–393

Becken, S. and Hay, J. (2007) *Tourism and Climate Change: Risks and Opportunities*, Channel View Publications, Clevedon and Buffalo, NY

Bedford, R. (2000) 'Meta-societies, Remittance Economies and Internet Addresses: Dimensions of Contemporary Human Security in Polynesia', in Graham, D. and Poku, N. (eds) *Migration, Globalisation and Human Security*, Routledge, London and New York

Bedford, R., Poot, J. and Ryan, T. (2006) *Niue: Population Policy Scoping Study:* Report for NZ Agency for International Development, University of Waikato, Hamilton

Beer, G. (1989) 'Discourses of the Island', in Amrine, F. (ed) *Literature and Science as Modes of Expression*, Kluwer, Dordrecht

Bell, J.D., Kronen, M., Vunisea, A., Nash, W.J., Keeble, G., Demmke, A., Pontifex, S. and Andréfouët, S. (2009) 'Planning the use of fish for food security in the Pacific', *Marine Policy*, vol 33, issue 1, pp64–76

Benedict, B. (1967) *Problems of Smaller Territories*, Athlone Press, London

Bertram, G. (1999) 'The MIRAB model twelve years on', *The Contemporary Pacific*, vol 11, issue 1, pp105–138

Bertram, G. and Watters, R. (1984) *New Zealand and its Small Island Neighbours*, Institute of Policy Studies, Wellington

Bertram, G. and Watters, R. (1985) 'The MIRAB economy in South Pacific microstates', *Pacific Viewpoint*, vol 26, issue 3, pp497–519

Black, R. (2001) 'Environmental refugees: Myth or reality?', Working Paper No.34 University of Sussex, UNHCR Evaluation and Policy Analysis Unit, Geneva

Boersma, P. D. (2008) 'Penguins as marine sentinels', *BioScience*, vol 58, issue 7, pp597–607

Böge, V. (1999) 'Mining, Environmental Degradation and War: The Bougainville Case', in Suliman, M. (ed) *Environment, Politics and Violent Conflict*, Zed Books, London and New York

Boland, S. and Dollery, B. (2007) 'The economic significance of migration and remittances in Tuvalu', *Pacific Economic Bulletin*, vol 22, issue 1, pp102–114

Bourke, R. M., Allen, M. G. and Salisbury, J. G. (eds) (2001) *Food Security for Papua New Guinea: Proceedings of the Papua New Guinea Food and Nutrition 2000 Conference*, PNG University of Technology, Lae, 26–30 June 2000, Australian Centre for International Agricultural Research, Canberra

Briguglio, L. (1995) 'Small island developing states and their economic vulnerabilities', *World Development*, vol 23, issue 9, pp1615–1632

Briguglio, L. (1997) *Alternative Economic Vulnerability Indices for Developing Countries*: Report prepared for the Expert Group on the Vulnerability Index, United Nations Department of Economic and Social Affairs, New York

Briguglio, L. and Kisanga, E. J. (eds) (2004) *Economic Vulnerability and Resilience of Small States*, Islands and Small States Institute, Malta and Commonwealth Secretariat, London

Briguglio, L., Cordina, G., and Kisanga, E. J. (eds) (2006) *Building the Economic Resilience of Small States*, Islands and Small States Institute, Malta and Commonwealth Secretariat, London

Brookfield, H. C. (1980) 'Introduction: The Conduct and Findings of the Interdisciplinary Fiji Project', in Brookfield, H.C. (ed) *Population-environment Relations in Tropical Islands: The Case of Eastern Fiji*, UNESCO, Paris

Brookfield, H. C. with Hart, D. (1971) *Melanesia. A Geographical Interpretation of an Island World*, Methuen, London

Brown, R. (1997) 'Estimating remittance functions for Pacific Island migrants', *World Development*, vol 25, issue 4, pp613–626

Browne, C. and Mineshima, A. (2007) 'Remittances in the Pacific Region', IMF Working Paper, IMF, Washington DC

Brunt, J., Hunter, D. and Delp, C. (2001) *A Bibliography of Taro Leaf Blight*, SPC, Noumea

Burton, I., Huq, S., Lim, B., Filifosova, O. and Schipper, L (2002) 'From impacts assessment to adaptation priorities: The shaping of adaptation policy', *Climate Policy*, vol 2, pp145–159

Burton, I., Kates, R.W. and White, G.F. (1993) *The Environment as Hazard*, 2nd Ed, Guilford Press, New York

Campbell, I. (1989) *A History of the Pacific Islands*, University of California Press, Berkeley

Campbell, J. R. (1985) *Dealing with Disaster. Hurricane Response in Fiji*, Pacific Islands Development Program, East–West Center, Honolulu

Campbell, J. R. (1990) 'Disasters and development in historical context: Tropical cyclone response in the Banks Islands, northern Vanuatu', *International Journal of Mass Emergencies and Disasters*, vol 8, issue 3, pp401–424

Campbell, J. R. (1997) 'Examining Pacific Island Vulnerability to Natural Hazards', in Planitz, A. and Chung, J. (eds) *Proceedings, VIII Pacific Science Inter-congress United Nations Department for Humanitarian Affairs*, South Pacific Programme Office, Suva

Campbell, J. R. (1998) 'Consolidating Mutual Assistance in Disaster Management Within the Pacific: Principles and Application', in South Pacific Applied Geoscience Commission (ed) *Seventh South Pacific Regional IDNDR Disaster Management Meeting*, South Pacific Applied Geoscience Commission, Suva

Campbell, J. R. (2006) *Traditional Disaster Reduction in Pacific Island Communities*: GNS Science Report, GNS Science, Wellington

Campbell, J. R. (2008) International Relocation from Pacific Island Countries: Adaptation Failure, Environment, Forced Migration & Social Vulnerability, International Conference, 9–11 October 2008, Bonn, Germany

Campbell, J. R. (2009) 'Islandness: Vulnerability and resilience in Oceania', *Shima: The International Journal of Research into Island Cultures*, vol 3, issue 1, pp85–97

Carruthers, P. (2002) 'Cook Islands Coastal Vulnerability Assessments: A Small Pacific Island Nation's Experience', available at www.survas.mdx.ac.uk/kobeproc.htm, accessed 6 May 2009

Carruthers, P. (2008) 'Gaps & Needs User Perspective Socioeconomic Information for Vulnerability and Adaptation Assessments', UNFCCC Expert Meeting under the Nairobi Work Programme on Adaptation, Port of Spain, 10 March

Castles, S. (2002) 'Environmental change and forced migration: Making sense of the debate', Working Paper No.70 University of Oxford Refugee Studies Centre, UNHCR Evaluation and Policy Analysis Unit, Geneva

Chambers, (1989) 'Editorial Introduction: Vulnerability, Coping and Policy', *IDS Bulletin*, vol 20, issue 2, pp 1–7

Chape, S. (2006) 'Review of Environmental Issues in the Pacific Region and the Role of the Pacific Regional Environment Programme', Workshop and Symposium on Collaboration for Sustainable Development of the Pacific Islands: Towards effective e-learning systems on environment, 27–28 February 2006, Okinawa, Japan

Chase, R. and Veitayaki, J. (1992) *Implications of Climate Change and Sea Level Rise for Western Samoa*, Report of a preparatory mission, SPREP, Apia

Chatterjee, P. and Finger, M. (1994) *The Earth Brokers: Power, Politics, and World Development*, Routledge, London

Clarke, W. C. (1977) 'The Structure of Permanence: The Relevance of Self-subsistence Communities for World Ecosystem Management', in Bayliss-Smith, T. and Feachem, R. (eds) *Subsistence and Survival: Rural Ecology in the Pacific*, Academic Press, London

Clarke, W. C., Manner, H. I. and Thaman, R. R. (1999) 'Agriculture and Forestry', in Rapaport, M. (ed) *The Pacific Islands Environment and Society*, Bess Press, Honolulu

CLIMsystems (n.d.) 'About CLIMsystems. Our Company', available at www.climsystems.com/about/, accessed 31 May 2009

Cocklin, C. and Keen, M. (2002) 'Urbanization in the Pacific: Environmental change, vulnerability and human security', *Environmental Conservation*, vol 27, issue 4, pp392–403

Commonwealth of Australia (2006) *Pacific 2020: Challenges and Opportunities for Growth*, Australian Agency for International Development, Canberra

Commonwealth of Australia (2008) *Pacific Economic Survey 08*, Australian Agency for International Development, Canberra

Commonwealth Secretariat (1997) *A Future for Small States: Overcoming Vulnerability*, Commonwealth Secretariat, London

Connell, J. (2003) 'Losing ground? Tuvalu, the greenhouse effect and the garbage can', *Asia Pacific Viewpoint*, vol 44, issue 2, pp89–107

Connell, J. (2006) ' "The taste of paradise": Selling Fiji and FIJI Water', *Asia Pacific Viewpoint*, vol 47, issue 3, pp342–350

Connell, J. and Lea, J. P. (2002) *Urbanisation in the Island Pacific: Towards Sustainable Development*, Routledge, London

Cutter, S. (1996) 'Vulnerability to environmental hazards', *Progress in Human Geography*, vol 20, issue 4, pp529–539

Dahan-Dalmedico, A. (2008) 'Climate expertise: Between scientific credibility and geopolitical imperatives', *Interdisciplinary Science Reviews*, vol 33, issue 1, pp71–81

Davis, W. J. (1996) 'The Alliance of Small Island States (AOSIS): The International Conscience', *Asia-Pacific Magazine*, vol 2, pp17–22

de Freitas, C. (2008) 'Tuvalu floods, but it's not sinking', *New Zealand Herald*, 19 March

Demeritt, D. (2001) 'The construction of global warming and the politics of science', *Annals of the Association of American Geographers*, vol 91, issue 2, pp307–333

Demeritt, D. (2006) 'Science studies, climate change and the prospects for constructivist critique', *Economy and Society*, vol 35, issue 3, pp453–479

Dessai, S., Adger, W., Hulme, M., Turnpenny, J., Köhler, J. and Warren, R. (2004) 'Defining and experiencing dangerous climate change', *Climatic Change*, vol 64, issue 1, pp11–25

Diamond, J. M. (2005) *Collapse: How Societies Choose to Fail or Succeed*, Viking, New York

DIVERSITAS (2007) *DIVERSITAS Annual Report 2006*, DIVERSITAS, Paris

Donner, S., Skirving, W., Little, C., Oppenheimer, M. and Hoegh-Guldberg, O. (2005) 'Global assessment of coral bleaching and required rates of adaptation under climate change', *Global Change Biology*, vol 11, issue 12, pp2251–2265

Dumaru, P. (2007a) *Climate Change Adaptation in Rural Communities of Fiji (CCA) Project: Progress Report July 2006–January 2007*, University of the South Pacific, Suva

Dumaru, P. (2007b) *Climate Change Adaptation in Rural Communities of Fiji (CCA) Project: Second Progress Report February–July 2007*, University of the South Pacific, Suva

Dumaru, P. (2008) *Climate Change Adaptation in Rural Communities of Fiji (CCA) Project: Third Progress Report, August–January 2007*, University of the South Pacific, Suva

Dumaru, P. (2009) 'Enhancing Capacity to Adapt to Climate Change in Four Fijian Villages: Understanding the Strengths and Limits of Community Based Adaptation', Seminar presented at the Department of Resource Management and Geography, University of Melbourne, Melbourne

Easterly, W. and Kraay, A. (2000) 'Small states, small problems? Income, growth, and volatility in small states', *World Development*, vol 28, issue 11, pp2013–2027

Eden, L. and Kudrie, R. (2005) 'Tax havens: Renegade states in the international tax regime?', *Law & Policy*, vol 27, issue 1, pp100–127

Eschenbach, W. (2004) 'Tuvalu not experiencing increased sea level rise', *Energy & Environment*, vol 15, issue 3, pp527–543

ESSP (Earth Systems Science Partnership) (2009) 'Homepage', available at www.essp.org/, accessed 11 April 2009

Fairbairn, P. (2008) 'Brief Historical Overview of the Round Table Meetings on Climate Change, Climate Variability and Sea-level Rise 2000–2008', Pacific Climate Change Roundtable, Samoa, 13–17 October

FAO (Food and Agriculture Organisation of the United Nations) (2008) 'Special Programme on Food Security', available at www.fao.org/spfs/ regional-programmes-rpfs/success-rpfs/pacific-islands/en/, accessed 24 November 2008

Farbotko, C. (2005) 'Tuvalu and climate change: Constructions of environmental displacement in the Sydney Morning Herald', *Geografiska Annaler*, vol87B, issue 4, pp279–293

Field, M. (2001) 'Time and Tide Holds No Favors For Tiny Tuvalu', available at http://archives.pireport.org/archive/2001/november/11-28-08.htm, accessed 20 June 2009

Field, M. (2002) 'Sinking or Rising, Future of Pacific Atolls All in the Entrails of Statistical Analysis', available at http://archives.pireport.org/archive/2002/august/08-27-05.htm, accessed 20 June 2009

Fisk, E. K. (1962) 'Planning in a primitive economy: Special problems of Papua New Guinea', *Economic Record*, vol 38, pp156–174

Flenley, J. R. and Bahn, P. G. (2003) *The Enigmas of Easter Island: Island on the Edge*, Oxford University Press, Oxford

Foucault, M. (1989) *The Archaeology of Knowledge*, London and New York, Routledge

Fraenkel, J. (2006) 'Beyond MIRAB: Do aid and remittances crowd out export growth in Pacific microeconomies?', *Asia-Pacific Viewpoint*, vol 47, issue 1, pp15–30

Fry, G. (1999) 'South Pacific security and global change: The new agenda', Working Paper No. 1999/1, Department of International Relations, RSPAS, Canberra

Fry, I. (2005) 'Small island developing states: Becalmed in a sea of soft law', *Review of European Community and International Environmental Law*, vol 14, issue 2, pp89–99

Funtowicz, S. and Ravetz, J. (1993) 'Science for the post-normal age', *Futures*, vol 25, issue 7, pp739–755

GEF (Global Environmental Facility) (2005) *GEF and Small Island Developing States*, GEF, Washington DC

GEF (2007) *Financing Adaptation Action*, GEF, Washington DC

GEF (2008) *Project Identification Form (PIF) GEFSEC Project ID: 3101. Project Title: Pacific Adaptation to Climate Change*, GEF, New York

Gibson-Graham, J.-K. (1998) 'Islands: Culture, Economy, Environment', in Bliss, E. (ed) *Conference Proceedings, Institute of Australian Geographers and New Zealand Geographical Society Conference, Hobart*, New Zealand Geographical Society, Palmerston North

Glantz, M. (1988) *Societal Responses to Regional Climate Change: Forecasting by Analogy*, Westview Press, Boulder, CO

Glantz, M. (2008) 'Tiempo interview: Mickey Glanz', *Tiempo*, vol 69, pp21–24

Government of Fiji (2004), 'Preliminary Estimates of Flood Affected Areas' Press Release, April 17, Suva

Government of Niue (2004), *National Impact Assessment Report of Cyclone Heta* Department of Economic, Planning, Development and Statistics, Alofi

Government of Tuvalu (2005) *Social Development Policy Project – Social Data Report 2005*, Community Affairs Department, Ministry of Home Affairs and Rural Development, Funafuti

Gowdy, J. and McDaniel, C. (1999) 'The physical destruction of Nauru: An example of weak sustainability', *Land Economics*, vol 75, issue 2, pp333–338

Gray, V. (2006) 'The truth about Tuvalu: A New Zealand climate scientist and a Pacific Island writer give assurances Tuvalu is not sinking', *New Zealand Climate and Environmental Truth*, available at http://nzclimatescience.net, accessed 20 June 2009

Grubb, M. with Vrolijk, C. and Brack, D. (1999) *The Kyoto Protocol: A Guide and Assessment*, Royal Institute of International Affairs, London

Haas, P. M. (2004) 'When does power listen to truth? A constructivist approach to the policy process', *Journal of European Public Policy*, vol 11, issue 4, pp569–592

Haitink, A. (1998) 'Small developing states and international organizations: The attention for SIDS in the works of the European Union, the United Nations and the Commonwealth, Background Paper'. Seminar on Small Island Developing States: Their Vulnerability, Their Programme of Action for Sustainable Development, Their Opportunities for Post-Lome, European Centre on Pacific Issues, Brussels, 1–2 September.

Hales, S., Weinstein, P. and Woodward, A. (1999) 'Ciguatera: Fish poisoning, El Niño, and sea surface temperature', *Ecosystem Health*, vol 5, pp5–20

Hall, P. (2006) 'What the south pacific sea level and climate monitoring project is telling us', Pacific Climate Change Discussions, AusAid, 10 November

Hall, P. (2007) 'South Pacific Sea Level & Climate Monitoring Project (SPSLCMP Phase IV)', UNFCCC Adaptation Workshop, Rarotonga, 26 February

Hampton, M. and Christensen, J. (2002) 'Offshore pariahs? Small island economies, tax havens, and the re-configuration of global finance', *World Development*, vol 30, issue 9, pp1657–1673

Hau'ofa, E. (1993) 'Our Sea of Islands', in Waddell, E. (ed) *A New Oceania: Rediscovering our Sea of Islands*, SSED, University of the South Pacific, Suva

Hau'ofa, E. (1994) 'Our sea of islands', *The Contemporary Pacific*, vol 6, pp147–161

Hay, J. (2000) *GEF Review of Climate Change Enabling Activities: Pacific Islands Regional Case Study*, GEF, Washington DC

Hay, J. and Sem, G. (2000) *Vulnerability and Adaptation: Evaluation and Regional Synthesis of National Assessments of Vulnerability and Adaptation to Climate Change*, SPREP, Apia

Hay, J., Mimura, N., Campbell, J., Fifita, S., Koshy, K., McLean, R. F., Nakalevu, T., Nunn, P. and de Wet, N. (2003) *Climate Variability and Change and Sea-level rise in the Pacific Islands Region: A Resource Book for Policy and Decision Makers, Educators and other Stakeholders*, SPREP, Apia

Hecht, A. and Tirpak, D. (1995) 'Framework agreement on climate change: A scientific and policy history', *Climatic Change*, vol 29, issue 4, pp371–402

Henderson, J. (2002) 'Pacific Freely Associated States: Seeking the Best of Both Worlds, *New Zealand International Review*, vol 27, issue 3, pp7–11

Henderson-Sellers, A. and Braaf, R. (1996) 'Developing New Perspectives on Climate Change, Impacts Assessment and Response', in Giambelluca, T. and Henderson-Sellers, A. (eds) *Climate Change: Developing Southern Hemisphere Perspectives*, John Wiley and Sons, Chichester

Hoegh-Guldberg, O., Hoegh-Guldberg, H., Stout, D., Cesar, H. and Timmerman, A. (2000) *Pacific in Peril: Biological, Economic and Social Impacts of Climate Change on Pacific Coral Reefs*, Greenpeace, Amsterdam

Holdgate, M. W., Bruce, J., Camacho, R. F., Desai, N., Mahtab, F. U., Mascarenhas, O., Maunder, W. J., Shihab, H. and Tewungwa, S. (1989) *Climate Change: Meeting the Challenge*, Report by a Commonwealth Group of Experts, Commonwealth Secretariat, London

Holdsworth, V. (2007) 'Climate change not only environmental', *Commonwealth Quarterly*, available at www.thecommonwealth.org/news/172515/161107climatechange.htm, accessed 5 July 2009

Holthus, P., Crawford, M., Makroro, C. and Sullivan, S. (1992) *Vulnerability Assessment of Accelerated Sea Level Rise, Case Study: Majuro Atoll, Marshall Islands*, SPREP, Apia

Hooper, A. (1983) 'Tokelau Fishing in Traditional and Modern Contexts', in Ruddle, K. and Johannes, R. E. (eds) *The Traditional Knowledge and Management of Coastal Systems in Asia and the Pacific*, UNESCO, Regional Office for Science and Technology for Southeast Asia, Jakarta

Hooper, A. (2000) 'Introduction', in Hooper, A. (ed) *Culture and Sustainable Development in the Pacific*, Asia Pacific Press, Canberra

Hughes, P. J. and McGregor, G. (eds) (1990) *Global Warming-Related Effects on Agriculture and Human Health and Comfort in the South Pacific*, SPREP, Noumea

Hulm, P. (1989) *A Climate of Crisis. Global Warming and the Island South Pacific*, ASPEI, Port Moresby

Hunter, J. R. (2002) *A Note on Relative Sea Level Change at Funafuti, Tuvalu*, Antarctic Cooperative Research Centre, Hobart

Hunter, J. R. (2004) 'Comments on "Tuvalu not experiencing sea-level rise" ', *Energy and Environment*, vol 15, issue 4, pp925–930

Hussein, B. (1997) 'The Big Retreat', *Pacific Islands Monthly*, Sydney, November, pp11–12

ICSU (International Council for Science)-IGFA (2008) *Review of the Earth System Science Partnership (ESSP)*, ICSU, Paris

Ielemia, A. (2008) Tuvalu Statement at The High Level Segment of the COP14 of the UNFCCC And CMP4 of the Kyoto Protocol Delivered By His Excellency The Honourable Apisai Ielemia Prime Minister And Minister Of Foreign Affairs, Fourteenth session of the Conference of the Parties (COP 14) and the fourth session of the Conference of the Parties serving as the meeting of the Parties to the Kyoto Protocol (COP/MOP 4), Poznan, 11 December 2008

IFRC (International Federation of Red Cross and Red Crescent Societies) (2003) *Preparedness for Climate Change: A Study to Assess the Future Impact of Climatic Changes Upon the Frequency and Severity of Disasters and the Implications for Humanitarian Response and Preparedness*, Netherlands Red Cross, The Hague

IFRC (2006) *Addressing the Humanitarian Consequences of Climate Change: Annual Report 2006*, Netherlands Red Cross, The Hague

IFRC (2008) 'Papua New Guinea: Cyclone Guba, Operations Update No.4', available at www.reliefweb.int/rw/RWFiles2008.nsf/FilesByRWDocUnidFilename/YSAR-7H4LHZ-full_report.pdf/$File/full_report.pdf, accessed 22 September 2008

IPCC (Intergovernmental Panel on Climate Change) (2004) *16 Years of Scientific Assessment in Support of the Climate Convention*, IPCC, Geneva

IPCC (2005) *IPCC Programme and Budget for 2006 to 2008*, IPCC Secretariat, Geneva

IPCC (2007) *Climate Change 2007: Synthesis Report*, Contribution of Working Groups I, II and III to the Fourth Assessment Report of the Intergovernmental Panel on Climate Change, IPCC, Geneva

Islands Business (2008) 'Poznan: We are no Lesser People, Pacific Tells World Talks', available at www.islandsbusiness.com/news/index_dynamic/containerNameToReplace=MiddleMiddle/focusModuleID=130/focusContentID=13926/tableName=mediaRelease/overideSkinName=newsArticle-full.tpl, accessed 5 July 2009

James, K. (1993) 'The rhetoric and reality of change and development in small Pacific communities', *Pacific Viewpoint*, vol 34, issue 2, pp135–152

Janssen, M., Schoon, M., Ke, W. and Börner, K. (2006) 'Scholarly networks on resilience, vulnerability and adaptation within the human dimensions of global environmental change', *Global Environmental Change*, vol 16, issue 3, pp240–252

Jones, R., Hennessy, K., Page, C., Walsh, K. and Whetton, P. (1999) *An Analysis of the Effects of the Kyoto Protocol on Pacific Island Countries: Regional Climate Change Scenarios and Risk Assessment Methods*, SPREP and CSIRO, Apia

Kaly, U., Pratt, C. and Howorth, R. (2002) 'A framework for managing environmental vulnerability in small island developing states', *Development Bulletin*, vol 58, pp22–38

Kaly, U., Pratt, C. and Mitchell, J. (2004) *The Environmental Vulnerability Index (EVI) 2004*, Technical Report 384, SOPAC, Suva

Kaly, U., Briguglio, L., McLeod, H., Schmall, S., Pratt, C., Schmall, S. and Pal, R. (1999) *Report on the Environmental Vulnerability Index (EVI) Think Tank*, Technical Report 299, SOPAC, Suva

Kaly, U., Pratt, C., Khaha, E., Dahl, A., Briguglio, L. and Sale-Mario, E. (2001) *Globalising the Environmental Vulnerability Index (EVI)*, Proceedings of the EVI Globalisation Meeting, Technical Report 345, SOPAC, Suva

Kaly, U., Pratt, C., Mitchell, J. and Howorth, R. (2003) *The Demonstration Environmental Vulnerability Index (EVI)*, Technical Report 356, SOPAC, Suva

Kandlikar, M. and Sagar, A. (1999) 'Climate change research and analysis in India: An integrated assessment of a South–North divide', *Global Environmental Change*, vol 9, issue 2, pp119–138

Kempf, W. (1999) 'Cosmologies, cities, and cultural constructions of space: Oceanic enlargements of the world', *Pacific Studies*, vol 22, issue 2, pp97–114

Kenny, G., Ye, W., Warrick, R. and de Wet, N. (2000) *FIJICLIM Description and Users Guide*, Report prepared for PICCAP by International Global Change Institute (IGCI), The University of Waikato, Hamilton

King, W. and Sem, G. (1999) *Pacific Islands Climate Change Assistance Programme, Programme Description*, SPREP, Apia

Kirch, P. V. (1997a) 'Microcosmic histories: Island perspectives on "global" change', *American Anthropologist*, vol 99, issue 1, pp30–42

Kirch, P. V. (1997b) 'Introduction: The Environmental History of Oceanic Islands', in Kirch, P. V. and Hunt, T. L. (eds) *Historical Ecology in the Pacific Islands: Prehistoric Environmental and Landscape Change*, Yale University Press, New Haven, CT

Kirch, P. V. (2000) *On The Road Of The Winds: An Archaeological History Of The Pacific Islands Before European Contact*, University of California Press, Berkeley, CA

Kiribati Government (1999) *Initial Communication Under The United Nations Framework Convention On Climate Change*, Ministry Of Environment And Social Development, Tarawa

Kiste, R. C. (1994) 'Pre-colonial Times', in Howe, K., Kiste, R. and Lal, B. (eds) *Tides of History: The Pacific Islands in the Twentieth Century*, University of Hawaii Press, Honolulu

Kleypas, J., Buddemeier, R., Archer, D., Gattuso, J., Landon, C. and Opdyke, B. (1999) 'Geochemical consequences of increased carbon dioxide in coral reefs', *Science*, vol 284, pp118–120

Komai, M. (2008) 'Poznan: Tuvalu's pleas receive global acknowledgement', *Islands Business*, available at www.islandsbusiness.com/news/index_dynamic/containerNameToReplace=MiddleMiddle/focusModuleID=130/

focusContentID=13960/tableName=mediaRelease/
overideSkinName=newsArticle-full.tpl, accessed 15 December 2008

Koshy, K. (2007) *Modeling Climate Change Impacts on Viti Levu (Fiji) and Aitutaki (Cook Islands)*, a Final Report submitted to AIACC Project No. SIS09, International START Secretariat, Washington DC

Kuletz, V. (2001) 'Invisible Spaces, Violent Places: Cold War Nuclear and Militarized Landscapes', in Peluso, N. and Watts, M. (eds) *Violent Environments*, Cornell University Press, Ithaca, NY

Lahsen, M. (2007) 'Trust Through Participation? Problems of Knowledge in Climate Decision Making', in Pettenger, M. (ed) *The Social Construction of Climate Change: Power, Knowledge, Norms, Discourses*, Ashgate, Aldershot

Lane, R. and McNaught, R. (2009) 'Building gendered approaches to adaptation in the Pacific', *Gender and Development*, vol 17, issue 1, pp67–80

Larmour, P. (1997) State Society and Governance in Melanesia, Corruption and Governance in the South Pacific: Discussion Paper 97/5, Research School of Pacific and Asian Studies, Australian National University, Canberra

Laupepa, P. (2002) 'High tides and low scientific standards', *Islands Business*, vol 28, issue 4, pp6–7

Leary, N. and Kulkarni, J. (2007) Climate Change Vulnerability and Adaptation in Developing Country Regions, Draft Final Report of the AIACC Project, International START Secretariat, Washington DC

Leary, N., Conde, C., Kulkarni, J., Nyong, A. and Pulhin, J. (eds) (2007a) *Climate Change and Vulnerability*, Earthscan, London

Leary, N., Adejuwon, J., Barros, V., Burton, I., Kulkarni, J. and Lasco, R. (eds) (2007b) *Climate Change and Adaptation*, Earthscan, London

Leiss, W. (1972) *The Domination of Nature*, George Braziller, New York

Lessa, W. (1964) 'The social effects of typhoon Ophelia (1960) on Ulithi', *Micronesia*, vol 1, pp1–47

Liberatore, A. (2001) 'From Arrhenius to the Kyoto Protocol: Climate Change and the Interplay Between Science and Policy', in Hisschemöller, M., Hoppe, R., Dunn, W. and Ravetz, J. (eds) *Knowledge, Power and Participation in Environmental Policy Analysis: Policy Studies Review Annual*, Transaction Publishers, New Jersey

Lieber, M.D. (ed.) (1977) *Exiles and Migrants in Oceania*. University Press of Hawaii, Honolulu.

Lim, B. and Spanger-Siegfried, E. (eds) (2004) *Adaptation Policy Frameworks for Climate Change: Developing Strategies, Policies and Measures*, Cambridge University Press, Cambridge

Luick, J. (2001) 'An Analysis of Variance in Pacific Tide Gauge Data', in Noye, J. and Grzechnik, M. (eds) *Sea-level Changes and Their Effects*, World Scientific Publishing, Singapore

MacArthur, N. and Yaxley, J. F. (1968) *Condominium of the New Hebrides: A Report on the First Census of the Population, 1967*, Condominium of the New Hebrides, Vila

MacArthur, R. and Wilson, E. (1967) *The Theory of Island Biogeography*. Princeton University Press, Princeton

Macdonald, G. (1983) 'Geology', in Armstrong, R. (ed) *Atlas of Hawaii*, 2nd edn, University of Hawaii Press, Honolulu

Mace, M. J. (2005) 'Funding for adaptation to climate change: UNFCCC and GEF Developments since COP-7', *Review of European Community & International Environmental Law*, vol 14, issue 3, pp225–246

Malnes, R. (2006) 'Imperfect science', *Global Environmental Politics*, vol 6, issue 3, pp58–71

Manner, H. I., Mueller-Dombois, D. and Rapaport, M. (1999) 'Terrestrial Ecosystems', in Rapaport, M. (ed) *The Pacific Islands Environment and Society*, Bess Press, Honolulu

Marland, G., Boden, T. and Andres, J. (2003) 'Global, Regional, and National CO2 Emissions', in Carbon Dioxide Information Analysis Center (ed) *Trends Online: A Compendium of Data on Global Change*, Oak Ridge National Laboratory, Oak Ridge, TN

Marshall, M. (1979) 'Natural and unnatural disaster in the Mortlock Islands of Micronesia', *Human Organization*, vol 38, issue 3, pp265–272

Mataki, M., Koshy, K and Lal, M (2006a) 'Baseline climatology of Viti-Levu (Fiji) and current climate trends', *Journal of Pacific Science*, vol 60, issue 1, pp49–68

Mataki, M, Koshy, K. and Nair, V. (2006b) 'Implementing climate change adaptation in the Pacific Islands: Adapting to present climate variability and extreme weather events in Navua (Fiji)', AIACC Working Paper 34, International START Secretariat, Washington DC

Mataki, M., Koshy, K. and Nair, V. (2007) 'Top-down, Bottom-up: Mainstreaming Adaptation in Pacific Island Townships', in Leary, N., Adejuwon, J., Barros, V., Burton, I., Kulkarni, J. and Lasco, R. (eds) *Climate Change and Adaptation*, Earthscan, London

McAdam, J. and Loughry, M. (2009) 'We Aren't Refugees', Inside Story (30 June 2009) http://inside.org.au/we-arent-refugees/ accessed 6 July, 2009

McFadzien, D., Areki, F., Biuvakadua, T. and Fiu, M. (2005) *Climate Witness Community Toolkit*, WWF South Pacific Programme, Suva

McGoldrick, W. (2009) 'Financing adaptation in Pacific Island countries: Prospects for the post-2012 climate change regime', *Australian Journal of International Law*, vol 14, pp45–69

McIntyre, M. (2005) *Pacific Environment Outlook*, UNEP and SPREP, Apia

McLean, R. and d'Aubert, A. (1993) *Implications of Climate Change and Sea Level Rise for Tokelau*, Report of a preparatory mission, SPREP, Apia

McNaught, R. (2007a) Pacific Red Cross Societies Preparing for a Changing Climate, available at www.climatecentre.org, accessed 16 June 2009

McNaught, R. (2007b) Small Island Developing States in the Pacific: Preparing for Climate Change, available at www.climatecentre.org, accessed 17 June 2009

McSaveney, M., Goff, J., Darby, D., Goldsmith, P., Barnett, A., Elliott, S. and Nongkas, M. (2000) 'The 17 July tsunami, Papua New Guinea: Evidence and initial interpretation', *Marine Geology*, vol 170, pp81–92

Mimura, N. (1999) 'Vulnerability of island countries in the South Pacific to sea level rise and climate change', *Climate Research*, vol 12, pp137–143

Mimura, N., Nurse, L., McLean, R. F., Agard, J., Briguglio, L., Lefale, P., Payet, R. and Sem, G. (2007) 'Small Islands', in Parry, M. L., Canziani, O. F., Palutikof, J. P., van der Linden P. J. and Hanson C. E. (eds) *Climate Change 2007: Impacts, Adaptation and Vulnerability*, Contribution of Working Group II to the Fourth Assessment Report of the Intergovernmental Panel on Climate Change, Cambridge University Press, Cambridge

Miranda, M., Burris, P., Bingcang, J. F., Shearman, P, Briones, J. O., La Viña, A. and Menard, S. (2003) *Mining and Critical Ecosystems: Mapping the Risks*, World Resources Institute, Washington DC

Mol, A. P. J. (2001) *Globalization and Environmental Reform: The Ecological Modernization of the Global Economy*, MIT Press, Cambridge, MA

Morgan, M. and Dowlatabadi, H. (1996) 'Learning from integrated assessment of climate change', *Climatic Change*, vol 34, issues 3–4, pp337–368

Morton, H. (2002) 'Creating Their Own Culture: Diasporic Tongans', in Spickard, P., Rondilla, J. and Wright, D. (eds) *Pacific Diaspora: Island Peoples in the United States and Across the Pacific*, University of Hawaii Press, Honolulu

Mortreux, C. and Barnett, J. (2009) 'Climate change, migration and adaptation in Funafuti, Tuvalu', *Global Environmental Change*, vol 19, issue 1, pp105–112

Moss, R. (1995) 'The IPCC: Policy relevant (not driven) scientific assessment', *Global Environmental Change*, vol 5, issue 3, pp171–174

Myers, N. (2002) 'Environmental refugees: A growing phenomenon of the 21st century', *Philosophical Transactions: Biological Sciences*, vol 357, pp609–613

Nakalevu, T. (2006) *Capacity Building for the Development of Adaptation Measures in Pacific Island Countries [CBDAMPIC] Project*, Final Report, SPREP, Apia

Nakalevu, T. (2007) 'Adaptation in Practice: The CBDAMPIC Project Approach', presented at the Stakeholder Workshop, Many Strong Voices: Climate Change Impacts and Adaptation in the Arctic and Small Island States, Belize, 27–30 May

Nakicenovic, N., Alcamo, J., Davis, G., de Vries, B., Fenhann, J., Gaffin, S., Gregory, K., Grubler, A., Jung, T., Kram, T., La Rovere, E., Michaelis, L., Mori, S., Morita, T., Pepper, W., Pitcher, H., Price, L., Riahi, K., Roehrl, A., Rogner, H., Sankovski, A., Schlesinger, M., Shukla, P., Smith, S., Swart, R., van Rooijen, S., Victor, N. and Dadi, Z. (2000) *Special Report on Emissions Scenarios*, Report by Working Group III of the Intergovernmental Panel on Climate Change, Cambridge University Press, Cambridge

Narsey, W. (2004) 'PICTA, PACER and EPAs: Weaknesses in Pacific Island countries' trade policies', *Pacific Economic Bulletin*, vol 19, issue 3, pp74–101

National Tidal Centre (2004) *The South Pacific Sea Level & Climate Monitoring Project (Phase III) Seaframe Station Calibration and Maintenance Program*, Report on Second Cycle, National Tidal Centre and Australian Bureau of Meteorology, Kent Town, South Australia

National Tidal Centre (2006) *The South Pacific Sea Level & Climate Monitoring Project*, Sea Level Data Summary Report, National Tidal Centre and Australian Bureau of Meteorology, Kent Town, South Australia

National Wildlife Federation (2008) 'Polar Bear Threatened Listing Weakened by Contradictions. Administration Still Running Away from Reality', available at http://www.nwf.org/news/ story.cfm?pageId=E9055F92%2DF1F6%2D7B10%2D3788A03291E5C483, accessed 13 December 2009

Nobel Foundation (2007) 'The Nobel Peace Prize 2007', available at http:// nobelprize.org/nobel_prizes/peace/laureates/2007/index.html, accessed 30 June 2009

Note, K. H., Remengesau, T. E. Jr and Urusemal, J. J. (2004) 'Joint Communiqué of the Fourth Micronesian Presidents' Summit, 19 July 2004', available at www.palaugov.net/PalauGov/Executive/thePRES/ jointcommunique.pdf, accessed 5 July 2009)

NRC (National Research Council) (2009) *Restructuring Federal Climate Research to Meet the Challenges of Climate Change*, Report of the Committee on Strategic Advice on the US Climate Change Science Program, The National Academies Press, Washington DC

Nunn, P. D. and Waddell, E. (1992) *Implications of Climate Change and Sea Level Rise for the Kingdom of Tonga*, SPREP, Apia

Nunn, P. D. (1994) *Oceanic Islands*, Blackwell, Oxford

Nunn, P. D. (2003a) 'AIACC Semi-Annual Reporting: January 15-July 15', available at http://sedac.ciesin.columbia.edu/aiacc/, accessed 11 May 2009

Nunn, P. D. (2003b) 'AIACC Project SIS09: Progress Report for July– December 2003', available at http://sedac.ciesin.columbia.edu/aiacc/, accessed 11 May 2009

Nunn, P. D. (2004) 'Understanding and Adapting to Sea-level Change', in Harris, F. (ed) *Global Environmental Issues*, Wiley, Chichester

Nunn, P. (2007) *Climate, Environment and Society in the Pacific During the Last Millennium*, Elsevier, Amsterdam

Nunn, P., Aalbersberg, W., Ravuvu, A. D., Mimura, N. and Yamada, K. (1994a) *Assessment of Coastal Vulnerability and Resilience to Sea-level Rise and Climate Change. Case Study: Yasawa Islands, Fiji. Phase 2: Development of Methodology*. Integrated Coastal Zone Management Programme for Western Samoa and Fiji, SPREP, Apia

Nunn, P, Balogh, E, Ravuvu, AD, Mimura, N and Yamada, K (1994b) *Assessment of Coastal Vulnerability and Resilience to Sea-level Rise and Climate Change. Case Study: Savai'i Island, Western Samoa. Phase 2: Development of Methodology*. Integrated Coastal Zone Management Programme for Western Samoa and Fiji, SPREP, Apia

Nunn, P. D., Aalbersberg, W., Clarke, C. C., Korovulayula, I., Mimura, N., Ohno, E., Yamada, K., Serizawa, M. and Nishioka, S. (1996) *Coastal Vulnerability and Resilience in Fiji. Assessment of Climate Change Impacts and Adaptation. Phase IV*. Integrated Coastal Zone Management Programme for Fiji and Tuvalu, SPREP, Apia

Nurse, L. and Moore, R. (2005) 'Adaptation to global climate change: An urgent requirement for small island developing states', *Review of European Community & International Environmental Law*, vol 14, issue 2, pp100–107

Nurse, L. and Sem, G. (2001) 'Small Island States', in McCarthy, J., Canziani, O., Leary, N., Dokken, D. and White, K. (eds) *Climate Change 2001: Impacts, Adaptation & Vulnerability*, Cambridge University Press, Cambridge

O'Brien, K. (2006) 'Are we missing the point? Global environmental change as an issue of human security', *Global Environmental Change*, vol 16, issue 1, pp1–3

O'Brien, K., Eriksen, S., Schjolden, A. and Nygaard, L. (2004) 'What's in a word? Conflicting interpretations of vulnerability in climate change research', CICERO Working Paper 2004:04, CICERO, Oslo

O'Brien, K., Barnett, J., De Scysa, I., Matthew, R., Mehta, L., Seager, J., Woodrow, M. and Bohle, H. (2005) 'Hurricane Katrina reveals challenges to human security', *AVISO*, vol 14, pp1–8

O'Collins, M. (1988) *Carteret Islanders at the Atolls Resettlement Scheme: A response to land loss and population growth. Potential impacts of greenhouse gas generated climatic change and projected sea-level rise on Pacific Island states of the SPREP Region*, ASPEI, Noumea

O'Collins, M. (1990) 'Carteret islanders at the Atolls Resettlement Scheme: a response to land loss and population growth', in Pernetta, J.C. and Hughes,P.J. (eds) *Implications of Expected Climate Changes in the South Pacific Region: an Overview*, UNEP Regional Seas Reports and Studies, UNEP, Nairobi.

O'Keefe, P., Westgate, K. and Wisner, B. (1976) 'Taking the naturalness out of natural disasters', *Nature*, vol 260, issue 5552, pp566–567

O'Riordan, T., Cooper, C. L., Jordan, A., Raynet, S., Richards, K. R., Runci, P. and Yoffe, S. (1998) 'Institutional Frameworks for Political Action', in Raynor, S. and Malone, E. L. (eds) *Human Choice and Climate Change. Volume 1: The Societal Framework*, Battelle Press, Columbus, OH

Oberthür, S. and Ott, H. (1999) *The Kyoto Protocol: International Climate Policy for the 21st Century*, Springer, Berlin

Olsen, K. (2007) 'The clean development mechanism's contribution to sustainable development: A review of the literature', *Climatic Change*, vol 84, issue 1, pp59–73

Olsthoorn, A., Maunder, W. and Tol, R. (1999) 'Tropical Cyclones in the Southwest Pacific: Impacts on Pacific Island Countries with Particular Reference to Fiji', in Downing, T., Olsthoorn, A. and Tol, R. (eds) *Climate, Change and Risk*, Routledge, London and New York

Overton, J. (1993) 'Small states, big issues? Human geography in the Pacific Islands', *Singapore Journal of Tropical Geography*, vol 14, issue 2, pp265–276

Palmer, G. (1988) The Greenhouse Effect and Its Relevance to the Pacific Region: East–West Center Lecture Series, East–West Center, Honolulu

Paeniu, B. (1991) 'Address by the Rt Hon. Bikenibeu Paeniu, Prime Minister of Tuvalu', in Jäger, J. and Ferguson, H. L. (eds) *Climate Change: Science, Impacts and Policy*, Proceedings of the Second World Climate Conference, Cambridge University Press, Cambridge

Palan, R. (2003) 'Tax havens and the commercialization of state sovereignty', *International Organization*, vol 56, issue 1, pp151–176

Parry, M., Palutikof, J., Hansen, C. and Lowe, J. (2008) 'Squaring up to reality', *Nature Climate Change Reports*, vol 2, pp68–70

Parson, E. and Fisher-Vanden, K. (1997) 'Integrated assessment models of global climate change', *Annual Review of Energy and the Environment*, vol 22, pp589–628

Paterson, M. (1996) *Global Warming and Global Politics*, Routledge, London

Pernetta, J. C. and Hughes, P. J. (eds) (1989) *Studies and Reviews of Greenhouse Related Climatic Change Impacts on the Pacific Islands*, SPC, Majuro

Pernetta, J. C. and Hughes, P. J. (eds) (1990) *Implications of Expected Climate Changes in the South Pacific Region: An Overview*, UNEP Regional Seas Reports and Studies, UNEP, Nairobi

Perry, M. (2005) 'Sinking Islands Cling to Kyoto Lifebuoy', *Planet Ark*, www.planetark.com/dailynewsstory.cfm/newsid/29499/story.htm accessed 13 December, 2009

Pettenger, M. (2007) 'Introduction: Power, Knowledge and the Social Construction of Climate Change', in Pettenger, M. (ed) *The Social Construction of Climate Change: Power, Knowledge, Norms, Discourses*, Ashgate, Aldershot

Pielke, R. (2005) 'Misdefining "climate change": Consequences for science and action', *Environmental Science and Policy*, vol 8, issue 6, pp548–561

Pielke, R. and Sarewitz, D. (2003) 'Wanted: Scientific leadership on climate', *Issues in Science and Technology*, Winter, pp27–30

PIFS (Pacific Islands Forum Secretariat) (2007) *The Pacific Plan for Strengthening Regional Cooperation and Integration*, PIFS, Suva

PIFS (2008) 'Forum Communiqué', Thirty-Ninth Pacific Islands Forum, Alofi, 19–20 August

Plumwood, V. (1993) *Feminism and the Mastery of Nature*, Routledge, London

Poirine, B. (1999) 'A theory of aid as trade with special reference to small islands', *Economic Development and Cultural Change*, vol 47, issue 4, pp831–852

Pratt, C., Kaly, U., Sale-Mario, E. and Seddon, J. (2002) *Towards a Global Environmental Vulnerability Index (EVI)*, Update on Progress March–June 2002 and Revised Funding Proposal 2002–2003, SOPAC, Suva

Proctor, J. (1998) 'The meaning of culture in global environmental change: Retheorizing culture in human dimensions research', *Global Environmental Change*, vol 8, issue 3, pp227–248

Radaelli, C. (1995) 'The role of knowledge in the policy process', *Journal of European Public Policy*, vol 2, issue 2, pp159–183

Rahmstorf, S. (2007) 'A semi-empirical approach to projecting future sea-level rise', *Science*, vol 315, issue 5810, pp368–370

Rainbird, P. (2002) 'A message for our future? The Rapa Nui (Easter Island) ecodisaster and Pacific Island environments', *World Archaeology*, vol 33, issue 3, pp436–451

Rappaport, R. (1968) *Pigs for the Ancestors*, Yale University Press, New Haven, CT

Ravuvu, A. D. (1987) *The Fijian Ethos*, Institute of Pacific Studies, University of the South Pacific, Suva

Ravuvu, A. D. (1988) *Development or Dependence: The Pattern of Change in a Fijian Village*, Institute of Pacific Studies and Fiji Extension Centre, University of the South Pacific, Suva

Reaser, J., Pomerance, R. and Thomas, P. (2000) 'Coral bleaching and global climate change: Scientific findings and policy recommendations', *Conservation Biology*, vol 14 pp1500–1511

Risbey, J., Kandlikar, M. and Patwardhan, A. (1996) 'Assessing integrated assessments', *Climatic Change*, vol 34, issues 3–4, pp369–395

Rogers, R. and Marres, N. (2000) 'Landscaping climate change: A mapping technique for understanding science and technology debates on the World Wide Web', *Public Understanding of Science*, vol 9, issue 2, pp141–163

Roncerel, A. (2002) *Lessons Learned from Using Modelling/Training Tools on Vulnerability and Adaptation Assessments in Developing Countries*, UNITAR, Bonn

Ronneberg, E. (2009) Outcome of the Pacific Climate Change Roundtable Process and Implications on the Energy Sector, Regional Energy Officials Meeting, Kingdom of Tonga, 20–22 April

Royle, S. A. (2001) *A Geography of Islands: Small Island Insularity*, Routledge, London

Runci, P. (2007) 'Expanding the participation of developing country scientists in international climate change research', *Environmental Practice*, vol 9, issue 4, pp225–227

Ruosteenoja, K., Carter, T., Jylha, K. and Tuomenvirta, H. (2003) *Future Climate in World Regions: An Intercomparison of Model-based Projections for the New IPCC Emissions Scenarios*, Finnish Environment Institute, Helsinki

Sabatier, P. (1988) 'An advocacy coalition framework of policy change and the role of policy-oriented learning therein', *Policy Sciences*, vol 29, issues 2–3, pp129–168

Sachs, W. (ed) (1993) *Global Ecology: A New Arena of Political Conflict*, Zed Books, London and New York

Said, E. W. (1978) *Orientalism*, Routledge and Kegan Paul, London

Sahlins, M. (1972) *Stone Age Economics*, Aldine de Gruyter, New York

Sahlins, M. (2000) 'On the Anthropology of Modernity, or Some Triumphs of Culture Over Despondency Theory', in Hooper, A. (ed) *Culture and Sustainable Development in the Pacific*, p48, Asia Pacific Press, Canberra

Salinger, J. (2001) 'Climate Variation in New Zealand and the Southwest Pacific', in Sturman, A. and Spronken-Smith, R. (eds) *The Physical Environment: A New Zealand Perspective*, Oxford University Press, Melbourne

Saloranta, T. (2001) 'Post-normal science and the global climate change issue', *Climatic Change*, vol 50, issue 4, pp395–404

Sampson, T. (2005) 'Aid to the Pacific: Past, present and future. Towards a new Pacific regionalism', Working Paper No. 18, an Asian Development Bank / Commonwealth Secretariat Joint Report to the Pacific Islands Forum Secretariat, ADB, Manila

Sarewitz, D. and Pielke, R. (2007) 'The neglected heart of science policy: Reconciling supply of and demand for science', *Environmental Science and Policy*, vol 10, issue 1, pp5–16

Schellnhuber, H., Cramer, W., Nakicenovic, N., Wigley, T. and Yohe, G. (eds) (2006) *Avoiding Dangerous Climate Change*, Cambridge University Press, Cambridge

Seager, J. (1993) *Earth follies: feminism, politics and the environment*. Routledge, New York.

Sem, G. and Underhill, Y. (1992) *Implications of Climate Change and Sea Level Rise for the Cook Islands*, Report of a preparatory mission, SPREP, Apia

Sem, G. and Underhill, Y. (1994) *Implications of Climate Change and Sea Level Rise for the Republic of Palau*, Report of a preparatory mission, SPREP, Apia

Sem, G., Mimura, N., Campbell, J. R., Hay, J. E., Ohno, E., Tamada, K., Serizawa, M. and Nishioka, S. (1996) *Coastal Vulnerability and Resilience in Tuvalu. Assessment of Climate Change Impacts and Adaptation. Phase IV*. Integrated Coastal Zone Management Programme for Fiji and Tuvalu, SPREP, Apia

Semple, E. C. (1911) *Influences of Geographic Environment on the Basis of Ratzel's System of Anthropo-Geography*, Constable, London

Sevele, F. (1987) 'Aid to the Pacific Reviewed', in Hooper, A., Britton, S., Crocombe, R., Huntsman, J. and Macpherson, C. (eds) *Class and Culture in the South Pacific*, University of Auckland and the University of the South Pacific, Auckland and Suva

Shackley, S. and Gough, C. (2002) 'The use of integrated assessment: An institutional analysis perspective', Working Paper 14, Tyndall Centre for Climate Change, Manchester

Shackley, S., Young, P., Parkinson, S. and Wynne, B. (1998) 'Uncertainty, complexity and concepts of good science in climate change modelling: Are GCMs the best tools?', *Climatic Change*, vol 38, issue 2, pp159–205

Sharma, K. L. (2006) *Food Security in the South Pacific Island Countries with Special Reference to the Fiji Islands*, Research Paper No. 2006/68, UNU-WIDER, Helsinki

Shibuya, E. (1996) 'Roaring mice against the tide: The South Pacific Islands and agenda building on global warming', *Pacific Affairs*, vol 69, issue 4, pp541–555

Shibuya, E. (2004) 'The Problems and Potential of the Pacific Islands Forum', in Rolfe, J. (ed) *The Asia-Pacific: A Region in Transition*, Asia-Pacific Center for Security Studies, Honolulu

Singh, R., Hales, S., de Wet, N., Raj, R., Hearnden, M. and Weinstein, P. (2001) 'The influence of climate variation and change on diarrheal disease in the Pacific Islands', *Environmental Health Perspectives*, vol 109, pp155–159

SMIC (Study of Man's Impact on Climate) (1971) *Inadvertent Climate Modification*, MIT Press, Cambridge, MA

SOPAC (South Pacific Applied Geoscience Commission) (2004) *Proceedings of the Thirty-third Session*, hosted by the Government of Papua New Guinea in the Coral Coast, Fiji Islands, 17–24 September

SOPAC (South Pacific Applied Geoscience Commission) (2005) EVI Country Profiles, available at www.vulnerabilityindex.net/EVI_Country_Profiles.htm, accessed 25 January 2006

SOPAC and UNEP (United Nations Environment Programme) (2005) *Building Resilience in SIDS: The Environmental Vulnerability Index*, SOPAC, Suva

Spary, E. and White, P. (2004) 'Food of paradise: Tahitian breadfruit and the autocritique of European consumption', *Endeavour*, vol 28, issue 2, pp75–80

SPC (Secretariat of the Pacific Community) (2004) *Pacific Islands Regional Millennium Development Goals Report 2004*, SPC, Noumea

SPC (2007) 'Agenda Item 6 – Climate Change: A Coordinated Organisational Response to Climate Change Issues', Meeting of the Conference of the Pacific Community, Apia, 12–13 November

SPC (2008a) *2008 Pocket Statistical Summary*, SPC, Noumea

SPC (2008b) *Pacific Island Populations 2008: 2008 Populations and Demographics Indicators*, SPC, Noumea

SPC (2008c) 'Agenda Item 3.2 – Climate Change: Contributions from SPC to Regional and National Adaptation', Thirty-eighth Meeting of the Committee Of Representatives of Governments and Administrations, Noumea, 13–16 October

SPREP (South Pacific Regional Environment Programme) (1992) *The Pacific Way: Pacific Island Developing Countries' Report to the United Nations Conference on Environment and Development*, SPREP, Noumea

SPREP (Secretariat of the Pacific Regional Environment Programme) (2005) *Pacific Islands Framework for Action on Climate Change 2006–2015*, SPREP, Apia

SPREP (2006) *Action Plan for the Implementation of the Pacific Islands Framework for Action on Climate Change 2006–2015*, SPREP, Apia

Spriggs, M. (1986) 'Landscape, Land Use, and Political Transformation in Southern Melanesia', in Kirch, P. (ed) *Island Societies: Archaeological Approaches to Evolution and Transformation*, Cambridge University Press, Cambridge

Spriggs, M. (1997) *The Island Melanesians*, Blackwell, Oxford

Stehr, N. and von Storch, H. (2005) 'Introduction to papers on mitigation and adaptation strategies for climate change: Protecting nature from society or protecting society from nature?', *Environmental Science and Policy*, vol 8, issue 6, pp537–540

Stern, N. (2007) *The Economics of Climate Change: The Stern Review*, Cambridge University Press, Cambridge

Sturman, A. and McGowan, H. A. (1999) 'Climate', in Rapaport, M. (ed) *The Pacific Islands Environment and Society*, Bess Press, Honolulu

Susman, P., O'Keefe, P. and Wisner, B. (1983) 'Global Disasters, a Radical Interpretation', in Hewitt, K. (ed) *Interpretations of Calamity from the Viewpoint of Human Ecology*, Allen and Unwin, Boston

Sutherland, K., Smit, B., Wulf, V. and Nakalevu, T. (2005) 'Vulnerability in Samoa', *Tiempo*, vol 54, pp11–15

Swyngedouw, E. (1999) 'Modernity and hybridity: Nature, regeneracionismo, and the production of the Spanish waterscape, 1890–1930', *Annals of the Association of American Geographers*, vol 89, issue 3, pp443–465

Teaiwa, K. (2005) 'Our Sea of Phosphate: The Diaspora of Ocean Island', in Harvey, G. and Thompson, C. (eds) *Indigenous Diasporas and Dislocations: Unsettling Western Fixations*, Ashgate, London

Thaman, R. R. (1994) 'Ethnobotany of Pacific Island Coastal Plants', in Morrison, J., Geraghty, P. and Crowl, L. (eds) *Science of Pacific Island Peoples*, Institute of Pacific Studies and University of the South Pacific, Suva

Thaman, R. R. (1995) 'Urban food gardening in the Pacific Islands: A basis for food security in rapidly urbanising small-island states', *Habitat International*, vol 19, issue 2, pp209–224

Thomas, W. L. (1963) 'The Variety of Physical Environments Among Pacific Islands', in Fosberg, F. R. (ed.) *Man's Place in the Island Ecosystem*, Bishop Museum Press, Honolulu

Tol, R. (2002) 'Estimates of the damage costs of climate change. Part 1: Benchmark estimates', *Environment and Resource Economics*, vol 21, issue 1, pp47–73

Tonn, B. (2007) 'The Intergovernmental Panel on Climate Change: A global scale transformative initiative', *Futures*, vol 39, issue 5, pp614–618

Torras, M. and Boyce, K. (1998) 'Income, inequality, and pollution: A reassessment of the environmental Kuznets Curve', *Ecological Economic*, vol 25, issue 2, pp147–160

Turnbull, J. (2001) Environmental Management In Fiji: A Socio-Political Exploration Of Global, Regional And State Dynamics, Thesis presented in partial fulfilment of the requirements for the degree of Master of Philosophy in Development Studies, Massey University, Palmerston North, New Zealand

Turnbull, J. (2003) 'South Pacific agendas in the quest to protect natural areas', *Development and Change*, vol 34, issue 1, pp1–24

Turner, S. and Holmes, G. (2001) 'Monitoring Sea Level: Who's Monitoring the Land?', in Noye, J. and Grzechnik, M. (eds) *Sea-level Changes and Their Effects*, World Scientific Publishing, Singapore

Turton, H. (2004) *Greenhouse Gas Emissions in Industrialised Countries: Where Does Australia Stand?*, The Australia Institute, Canberra

Tuvalu Red Cross Society (2008) Tuvalu: Joining Forces to Tackle Climate Change, available at www.ifrc.org/Docs/pubs/disasters/resources/corner/case-studies/cs-tuvalu-en.pdf, accessed 17 June 2009

UNEP (United Nations Environment Programme) (2008) *Kick the Habit: A UN Guide to Climate Neutrality*, UNEP, Nairobi

UNEP/WMO/ICSU International Conference (1986) 'Statement by the UNEP/WMO/ICSU International Conference on the Assessment of the Role of Carbon Dioxide and of Other Greenhouse Gases in Climate Variations and Associated Impacts, Villach, Austria, 9–15 October 1985', in Bolin, B., Döös, B. R., Jäger, J., and Warrick, R. A. (eds) *The Greenhouse Effect, Climatic Change, and Ecosystems*, Wiley, New York

United Nations (1994) *Report of the Global Conference on the Sustainable Development of Small Island Developing States*, Bridgetown, Barbados, 25 April to 6 May 1994, United Nations Department of Public Information, New York

UNFCCC (United Nations Framework Convention on Climate Change) (2002) *Annotated Guidelines for the Preparation of National Adaptation Programmes of Action*, LDC Expert Group, UNFCCC Secretariat, Bonn

UNFCCC (2005) *Climate Change, Small Island Developing States.* Climate Change Secretariat (UNFCCC), Bonn

UNFCCC (2008) *Conference of the Parties, Fourteenth Session, Poznan, 1–12 December 2008 List of Participants, Part 1*, UNFCCC, Geneva

United Nations News Centre (2008) ' "Small Island Nations" Survival Threatened by Climate Change, UN Hears', available at www.un.org/apps/news/story.asp?NewsID=28265&Cr=general+assembly&Cr1=debate, accessed 5 July 2009

Vidal de la Blache, P. (1926) *Principles of Human Geography*, Constable, London

Villa, F. and McLeod, H. (2002) 'Environmental vulnerability indicators for environmental planning and decision-making: Guidelines and applications', *Environmental Management*, vol 29, pp335–348

Vogler, J. and Bretherton, C. (2006) 'The European Union as a protagonist to the United States on climate change', *International Studies Perspectives*, vol 7, issue 1, pp1–22

Waddell, E. (1975) 'How the Enga coped with frost: Responses to climatic perturbations in the Central Highlands of New Guinea', *Human Ecology*, vol 4, issue 4, pp249–273

Walton, G. and Barnett, J. (2008) 'The ambiguities of environmental degradation and violence: The case of the Tolukuma gold mine', *Society and Natural Resources*, vol 21, issue 1, pp1–16

Warrick, R., Kenny, G., Sims, G., Ye, W. and Sem, G. (1999) 'The Vandaclim Simulation Model: A training tool for climate change vulnerability and adaptation assessment', *Environment, Development and Sustainability*, vol 1, issue 2, pp157–170

Watters, R. (1987) 'Mirab Societies and Bureaucratic Elites', in Hooper, A., Britton, S., Crocombe, R., Huntsman, J. and Macpherson, C. (eds) *Class and Culture in the South Pacific*, University of Auckland and University of the South Pacific, Auckland and Suva

Watts, M. (1983) 'On the Poverty of Theory: Natural Hazards Research in Context', in Hewitt, K. (ed) *Interpretations of Calamity from the Viewpoint of Human Ecology*, Allen and Unwin, Boston, MA

Watts, M. (1993) 'Hunger, famine and the space of vulnerability', *GeoJournal*, vol 30, issue 2, pp117–125

Watts, M., and Bohle, H. G. (1993) 'The space of vulnerability: The causal structure of hunger and famine', *Progress in Human Geography*, vol 17, issue 1, pp43–67

WCED (World Commssion on Environment and Development) (1987) *Our Common Future*, Oxford University Press, Oxford

WCRP (World Climate Research Programme) (2008) *WCRP Accomplishment Report 2007–2008*, WCRP, Geneva

Weart, S. R. (2008) *The Discovery of Global Warming*, Harvard University Press, Cambridge, MA

Webb, A. (2006) *Tuvalu Technical Report: Coastal Change Analysis Using Multi-Temporal Image Comparisons, Funafuti Atoll*, SOPAC Project Report 54, SOPAC, Suva

Webb, J. (2008) *Engaging Young People in the Solomon Islands for Red Cross Action on Climate Change*, The Climate Centre, Netherlands

White, D. F. (2006) 'A political sociology of socionatures: Revisionist manoeuvres in environmental sociology', *Environmental Politics*, vol 15, issue 1, pp59–77

Wilson, M. (2001) 'Location location location: The geography of the dot com problem', *Environment and Planning B: Planning and Design*, vol 28, issue 1, pp59–71

Wisner, B., Blaikie, P., Cannon, T. and Davis, I. (2004) *At Risk. Natural Hazards, People's Vulnerability and Disasters*, Routledge, London

Withers, C. W. J. (1999) 'Geography, Enlightenment, and the Paradise Question', in Livingstone, D. N., and Withers, C. W. J. (eds) *Geography and Enlightenment*, University of Chicago Press, Chicago, IL

WMO (World Meteorological Organization) (1986) 'Report of the International Conference on the assessment of the role of carbon dioxide and of other greenhouse gases in climate variations and associated impacts, Villach, Austria', available at www.icsu-scope.org/downloadpubs/scope29/statement.html, accessed 5 July 2009

WMO (1999) *The 1997–1998 El Niño Event: A Scientific and Technical Retrospective*, WMO, Geneva

WMO, UNEP (United Nations Environment Programmes) and Environment Canada (1989) 'Conference Proceedings', *The Changing Atmosphere: Implications for Global Security*, Toronto, 27–30 June

World Bank (2000) *Cities, Sea, and Storms: Managing Climate in Pacific Island Economies, Vol. IV, Adapting to Climate Change*, The World Bank, Washington DC

World Resources Institute (2008) Climate Analysis Indicators Tool (CAIT), available at http://cait.wri.org, accessed 22 December. 2008

Yamada, K., Nunn, P. D., Mimura, N., Machida, S. and Yamamoto, K. (1995) 'Methodology for the assessment of vulnerability of South Pacific island countries to sea-level rise and climate change', *Journal of Global Environment Engineering*, vol 1, pp101–125

Yamano, H., Kayanne, H., Yamaguchi, T., Kuwahara, Y., Yokoki, H., Shimazaki, H. and Chikamori, M. (2007) 'Atoll island vulnerability to flooding and inundation revealed by historical reconstruction: Fongafale Islet, Funafuti Atoll, Tuvalu', *Global and Planetary Change*, vol 57, issues 3–4, pp407–416

Index